# Air Pollution Effects on Plant Growth

# Air Pollution Effects on Plant Growth

Mack Dugger, *Editor*

A symposium sponsored by
the Division of Agricultural
and Food Chemistry at the
167th Meeting of the American
Chemical Society,
Los Angeles, Calif.,
April 1, 1974.

ACS SYMPOSIUM SERIES **3**

AMERICAN CHEMICAL SOCIETY

WASHINGTON, D. C.     1974

Library of Congress CIP Data

Air pollution effects on plant growth.
(ACS symposium series; 3)

Includes bibliographical references and index.

1. Plants, Effect of air pollution on—Congresses.
I. Dugger, Mack, 1919- ed. II. American Chemical
Society. Division of Agricultural and Food Chemistry.
III. Series: American Chemical Society. ACS symposium
series; 3. [DNLM: 1. Air pollution—Congresses. 2.
Plant diseases—Congresses. SB745 A298 1974]

QK751.A37          582'.05'222          74-26543
ISBN 0-8412-0223-0          ACSMC8 3 1-150 (1974)

# ACS Symposium Series

Robert F. Gould, *Series Editor*

# FOREWORD

The ACS SYMPOSIUM SERIES was founded in 1974 to provide a medium for publishing symposia quickly in book form. The format of the SERIES parallels that of its predecessor, ADVANCES IN CHEMISTRY SERIES, except that in order to save time the papers are not typeset but are reproduced as they are submitted by the authors in camera-ready form. As a further means of saving time, the papers are not edited or reviewed except by the symposium chairman, who becomes editor of the book. Papers published in the ACS SYMPOSIUM SERIES are original contributions not published elsewhere in whole or major part and include reports of research as well as reviews since symposia may embrace both types of presentation.

# CONTENTS

# PREFACE

Twenty-five to thirty years ago widespread oxidant air pollution damage to ornamental and crop plants became apparent to both scientists and the public. This occurred first in the Los Angeles area, and it was predominantly associated with the large increase in industrial activity in the Los Angeles basin during the 1940's and 1950's. Research workers in and near Los Angeles, particularly at the California Institute of Technology at Pasadena and the University of California Agricultural Experiment Station at Riverside, conducted investigations to determine the cause and effect relationship between observed plant damage and environmental contaminants. Significant efforts were made to understand the complex atmospheric chemical and photochemical reactions leading to formulation of air pollutants identified in the atmosphere of the Los Angeles Basin. In later development of oxidant air pollution knowledge, automobile exhaust emissions and other fuel combustion processes were implicated as major factors in producing this class of air pollutants both in and near large metropolitan areas. In time, it became apparent that other areas in this country and throughout the world were experiencing similar oxidant air pollution episodes.

Some of the pioneers in these early efforts to identify oxidants and understand the mechanisms of how they damage plants were A. J. Haagen-Smit and F. W. Went and their colleagues at the California Institute of Technology, and John T. Middleton, Ellis F. Darley, and J. B. Kendrick, Jr. and their colleagues at the University of California Agricultural Experiment Station at Riverside. Substantial progress was made throughout the beginning years of air pollution research by these southern California researchers. Other research teams throughout the United States were also established to investigate plant responses to air pollution. Several of the major laboratories conducting early work in this research area were the USDA Laboratory at Beltsville, Md.; the Connecticut Agriculture Experiment Station at Windsor, Conn.; the National Center for Air Pollution Control, U. S. Department of Health, Education, and Welfare at Cincinnati, Oh.; and the University of Utah, Utah State University, Pennsylvania State University, and Rutgers University. During the 1950's and 1960's, significant advancement was made in determining chemical, meteorological, and biological knowledge regarding atmospheric oxidant damage to biological systems.

Within the past few years, the rate of our understanding of how air pollution oxidants interect with biological systems has accelerated dramatically. This is the result partially of a reasonable level of federal grants and contract funds and partially of cooperative support by local and state agencies of research into specific pollution problems. Research supported by these agencies not only has helped establish an air pollution control policy, it has also helped develop a better scientific understanding of the biological effects of air pollution on man and plants.

It was appropriate, therefore, that the Division of Agricultural and Food Chemistry hold a symposium on the topic of air pollution and equally appropriate that this meeting be held in Los Angeles. The papers in this volume describe not only some of the most recent ideas concerning this topic, but they also describe tests of these ideas by researchers who are second and third generation scientists in this field. In a sense, the questions asked by these more recent scientists do not differ significantly from those posed by the pioneers in air pollution studies; however, this does not imply that no progress has been made. As is the case with any area of scientific inquiry, questions often remain the same while answers change and become increasingly more complex. In this particular case, the complexity of the answers is attributable not only to the accumulation of new knowledge, but also to how critically important that new knowledge is in its potential and actual effects on man and his environment. Future symposia on this subject will disclose still other answers to as yet unresolved questions. In time, however, we should understand much more completely the cause and effect relationship between oxidant air pollution and the mechanism of plant damage.

The major photochemically-produced atmospheric pollutant is ozone and, accordingly, most of the papers in this volume are concerned with this oxidant and its effects on plant processes, plant constituents such as cellular membranes, or on biochemical reactions common to the biological systems of plants and animals. I believe these authors have significant new answers to old questions, and I applaud them for their contributions.

I would like to thank Michael Elderman for his invaluable editorial assistance in the preparation of these manuscripts.

MACK DUGGER

Riverside, Calif.
September 20, 1974

# Air Pollutant Effects Influenced by Plant–Environmental Interaction

O. C. TAYLOR

Statewide Air Pollution Research Center, University of California,
Riverside, Calif. 92502

Susceptibility of vegetation to air pollutant injury is re-
portedly influenced by many climatic factors, edaphic factors,
genetic variability, and by structural and biochemical variations
in the plant. Observed differences in susceptibility have
prompted numerous attempts to prepare lists of plant species
according to susceptibility or tolerance to specific air pollutants. It has become apparent that such lists are frequently
inaccurate because the differential in susceptibility within a
single variety or cultivar may be as great as that between
species.

Accuracy of tolerance lists is further complicated by the
fact that data are often available from only one locality or
from a single experimental laboratory where one set of environmental factors interact with the pollutant reaction. Differences
in plant age, variety, rate of growth, light exposure, climatic
conditions, nutrition and other factors can induce significant
variation in response to a particular pollutant.

The complex interaction between climatic factors, toxicant
concentration, exposure duration, soil conditions and physiological characteristics determines the susceptibility of plant
tissue and, to a large extent, influences the characteristics
of the injury symptom syndrome produced. Distinct and relatively uniform symptoms may be produced repeatedly on susceptible
members of the plant community if they are exposed under
controlled-standardized conditions. But the extent of injury and
a concomitant alteration in injury appearance usually occurs when
there is a deviation in toxicant dosage, plant species or environmental condition.

## Meteorological Factors

Light. Photoperiod has a marked effect on susceptibility of
plants to photochemical air pollutants. The effect of photoperiod is particularly striking when peroxyacetyl nitrate (PAN)
is the principal toxicant, but there is also a significant effect

when ozone and nitrogen dioxide are used as phytotoxicants.
Juhren et al. (1) reported greater susceptibility of annual blue-
grass to oxidant with an 8-hr photoperiod than with a 16-hr
photoperiod, where plants were grown under moderate light intensity.  Heck and Dunning (2) and MacDowall (3) also found that
plants growing under a short day (8 hr.) were more susceptible
than plants growing under a 12 or 16-hr/day regime.  Generally,
plants growing continuously under short-day conditions are most
susceptible to oxidants, but MacDowall (3) showed that susceptibility of plants growing under long-day conditions could be increased by a 3-day treatment with 8 hr of light prior to fumigation.

     Taylor et al. (4) suggested that interactions between light
and oxidants from the polluted atmosphere within plant tissues
might explain not only some of the variabilities in symptoma-
tology observed in controlled experiments but also the unex-
plained variability observed in some polluted areas.  Why did
exposure to PAN on one day fail to produce injury symptoms, and
an exposure of the same concentration and duration on another day
produce severe injury?  Further investigation revealed that
susceptibility of pinto bean and petunia plants to PAN was significantly affected by the presence of light immediately before,
during and after exposure to the toxicant (4, 5).

     The light-dark interactions which regulated susceptibility
of bean plants to PAN injury were described in detail by Taylor
(6).  Approximately 2.75 hr of sunlight following a 12-hr night
was required for full susceptibility to develop; a 15 min dark
period before a 30-min fumigation was sufficient to give complete
protection.  A full hour of sunlight following the 15-min dark
treatment was required to regain complete susceptibility.  It is
important to recognize the effect of short, dark exposures to
avoid errors in interpreting response during fumigation studies.
A brief stay in darkness or very low light intensity may be
sufficient to provide a significant amount of tolerance to PAN.

     Taylor (6) also reported that continued exposure to sunlight
for about 3 hr, following a 30-min fumigation, was required for
complete development of PAN symptoms.  Plants placed in a dark
chamber immediately after a 30-min PAN treatment were not injured.  The delayed effect of PAN was also observed in the
Riverside laboratory when an attempt was made to measure changes
in $CO_2$ absorption during and following PAN treatment.  Absorption
rate, measured in a dynamic flow system with a nondispersive in-
frared analyzer, did not change during a one-hr fumigation period
or during 2.75 hr immediately following fumigation.  After about
2.75 hr, water-soaked areas became faintly visible, and $CO_2$
absorption declined abruptly to approximately 50% of the former
level.  Full development of the lower surface bronzing and
collapsed "bifacial" lesions occurred during the succeeding 36 hr.

     Air pollutants are transported with wind movement; conse-
quently, elevated concentrations may be moved from one area to

another.  Peak concentrations of ozone and PAN in the eastern
part of the South Coast Air Basin of California usually occur in
the afternoon and evening.  If the polluted air mass reaches
vegetation late in the evening, PAN injury will probably not
develop even though the threshold concentration is exceeded
through the night (6).  Severe injury can be expected if the
pollutants reach the area slightly earlier in the day.

Photoperiod also affects the response of plants to ozone
(7, 8).  Ting and Dugger (8) reported that after a 24-hr dark
treatment, leaves of cotton plants were no longer sensitive to
ozone.  He pointed out that radiant energy during development
correlated with susceptibility of cotton to ozone.  Davis (7)
found that seedlings of Virginia pine were protected from injury
when they were kept in light for 24 hr or longer prior to exposure to damaging doses of ozone.  They also found that pine
seedlings kept in darkness for up to 96 hr prior to ozone exposure were susceptible to injury.  These conflicting experi-
ences with the protective effect of pre-fumigation dark treatments suggest that further studies may be beneficial.  Dugger
et al. (5) found that susceptibility of young bean plants to
ozone was increased by dark periods of up to 24 hr, but the
tolerance increased dramatically with longer dark periods.

Heck (9, 10) and his co-workers noted that Juhren et al.
(12) reported Poa annua plants were most susceptible to Los
Angeles type smog when they were grown in a short (8-hr) photoperiod and that Dugger et al. (5) reported that bean plants
grown under 900-ft-c light intensity were more susceptible than
plants grown under 2200-ft-c.  Consequently, they investigated
combinations of photoperiod length and light intensity and found
greatest susceptibility of bean and tobacco plants to ozone
occurred when they were grown under low light intensity with
short photoperiod.  Observations have also been made that extensive areas of white or light tan necrotic lesions of collapsed
tissue are common when plants are exposed to ozone under low
light intensity.  This bleaching effect is less pronounced and
the necrotic lesions appear dark brown or reddish-brown when
plants are exposed under full sunlight.  An optimum level of
sugar (sucrose and reducing sugar) in leaf tissue was required to
induce maximum susceptibility (11, 12, 13).  Lee (13) suggested
that the sugar content regulated stomatal opening and thus had
a marked effect on the amount of injury.  Dugger's group (11, 12)
postulated that the sugar relation to tolerance was more direct.

**Temperature.**  Hill et al. (14) initiated studies of effects
of sulfur dioxide and mixtures of sulfur dioxide and nitrogen
dioxide on desert vegetation by transplanting native species to
the greenhouse, but they later concluded that sensitivity varied
greatly with differences in soil moisture, relative humidity and
temperature.  Consequently, fumigations were carried out in the
field with plants growing under natural conditions.  Davis (7)

observed an inverse relationship between exposure temperature
and degree of injury on Virginia pine. Conversely, Cameron and
Taylor (15) observed that elevated concentrations of ozone in
the field produced little or no visible symptoms in sweet corn
if ambient air temperatures during and after exposures were below
about 90°F. Severe injury occurred on susceptible varieties
of sweet corn when temperature exceeded 90°F. Davis (7) observed a direct correlation between the amount of ozone injury
produced and the temperature at which pine seedlings were maintained before and after exposures.

Relative Humidity. Heggestad et al. (16) suggested that the
high relative humidity along the east coast of the U. S. increased susceptibility of plants to ozone injury. The tobacco
growing spaces and fumigation chambers used in his investigations
were designed to maintain saturated atmosphere during all stages
of growing and fumigation of plants. The greater amount of ozone
required to produce ozone injury in west coast fumigations was
attributed to the low relative humidity in that area. Efforts
to prove this concept by using fog nozzles in the plant growing
areas and the use of fog nozzles to add water to the air intake
of fumigation chambers used at Riverside, California did not
conclusively confirm the theory. Increased tolerance of plants
to ozone has been observed when fumigations were conducted in air
drier than that in the growing area. When plants were trans-
ferred from the greenhouse to fumigation chambers through the
dry ambient air at Riverside, stomatal closure due to the immediate water stress was considered to be the reason for the pronounced increase of tolerance under such conditions. Davis (7)
found that pine seedlings developed more severe ozone injury
when exposed in high humidity than when exposed in low humid-
ities. In their studies, humidity before and after exposure
to 25 pphm ozone for 4 hr had no effect upon the amount of injury
produced. MacDowell (17) reported that tobacco was more susceptible to ozone when dew was on the leaves than when they were
dry.

There is no conclusive experimental evidence that relative
humidity significantly affects susceptibility of plants to PAN
as long as conditions are such that the stomata remain open.
There have been suggestions, however, that high relative humidity
may cause rapid breakdown of PAN (18) but there is no experimental evidence to support such a statement. In the South Coast
Air Basin of California, PAN injury to vegetation occurs most
frequently when relative humidity is 50% or above and elevated
concentrations of 10 to 30 ppb often persist all night, with
relative humidity above 60%.

Edaphic Factors. Soil texture, soil moisture and mineral
nutrients strongly influence plant growth and, consequently, have
an effect on susceptibility to air pollutants. Adequate soil
moisture to maintain leaf turgidity is essential to maintain full

susceptibility of plants.  Field observations in southern
California, where irrigation is essential for crop production,
have revealed that crops grown under soil moisture deficit
developed little or no ozone or PAN type symptoms during a severe
smog attack, while adjacent recently-irrigated crops were
severely injured.  Withholding irrigation has been suggested as a
technique for preventing oxidant air pollutant damage (19).
One might be fortunate if a smog attack of a single day's duration should coincide with a period when the field is dry, but the
damage caused by withholding water in anticipation of a smog
attack could be as injurious to the crop as the air pollutant.

Seidman et al. (20) found that water could be withheld from
petunia, tobacco and pinto bean to the point just short of wilting to close stomata and prevent injury from irradiated automobile exhaust.  MacDowall (3) reported that tobacco grown for
several days with deficient soil moisture, but just short of
wilting, were less susceptible to ozone than normal plants even
though they were well watered just prior to exposure.  Similarly,
increased tolerance of beans and tobacco to ozone and PAN has
been noted when test plants were inadvertantly allowed to wilt
briefly during the day preceding fumigation, even though they
were well watered several hours before treatment and appeared
to be normal.

Excessive soil water for an extended period may reduce
susceptibility of plants to ozone injury.  Stolzy et al. (21)
showed that oxygen deficiency in the root zone increased ozone
tolerance of tomato plants.  The low partial pressure of oxygen
in the root zone also reduced water uptake and plant vigor.
Seidman et al. (20) found that plant vigor and susceptibility to
ozone were less with plants grown in clay soils rather than in
vermiculite; similarly, vigor and susceptibility of plants grown
in clay loam was suppressed compared to plants grown in a peat-
perlite mix.

Nutrition.  Middleton (19) reported that spinach and lettuce
were more susceptible to ozonated hexene when they received an
adequate supply of nitrogen fertilizer.  Others have also reported increased susceptibility to oxidants when nitrogen fertilizer was applied (22, 23).  There have been some reports that
indicated air pollutant injury was greatest in tests with the
least amount of nitrogen added (3, 10, 24).  This suggests that
injury was enhanced by the addition of nitrogen when there was
a deficiency but a luxury amount did not increase injury and perhaps even suppressed growth to the point of inducing greater
tolerance.

Oertli (25) reported increased tolerance of sunflower plants
to oxidant air pollutants (smog) with increasing salinity and
soil moisture stress.  However, Heck et al. (9, 10) suggested
that Oertli was dealing with a nutrient as well as osmotic
effect.

Mixtures of Toxicants. Polluted atmospheres are complex
mixtures of many toxic and inert materials which may interact to
increase or diminish plant injury. Menser and Heggestad (26)
reported the first evidence of synergistic response when an
ozone-susceptible tobacco strain was exposed to a mixture of ozone
and sulfur dioxide. Other researchers have subsequently confirmed
this response, but attempts to identify synergistic or antagonistic responses with other pollutants have been unsuccessful,
although synergistic effects of $NO_2$ and $SO_2$ are suggested.
Further study of the possible interaction of pollutants would be
desirable to help explain variations in plant response in
different locations.

Discussion and Summary. Genetic variability is a major
determinant of plant tolerance to air pollutants. Differences
in susceptibility between species or genera are well known, but
significant variation between individuals within a variety or
cultivar has been less obvious. Tolerance and characteristics of
symptoms are also dependent upon the way exposure to the pollutants occur. Exposure to high concentrations of a toxic pollutant
usually produces symptoms quite different from those produced by
long exposure to low concentrations. Heck and Dunning (2, 10)
reported that plant susceptibility was greater when exposed to
a given dose of ozone in one hr than when the same dose was
applied in multiple exposures.

A number of external and internal factors are influential in
determining the susceptibility of a specific plant to an airborne
toxicant. Some generalizations can be stated relative to factors
that affect susceptibility to oxidant air pollutants:

    a. Rapid-vigorous growth usually increased susceptibility.
    b. Plants grown in intense light are more susceptible to
       PAN.
    c. Low light intensity enhanced susceptibility to ozone.
    d. A long dark period (24 hr) prior to exposure increased
       ozone susceptibility.
    e. Light before and after exposure to PAN is required for
       injury development.
    f. A long photoperiod (24 hr) before exposure inhibited
       ozone injury.
    g. High temperature during exposure in the field increased
       ozone susceptibility.
    h. High relative humidity increased susceptibility to ozone
       and PAN.
    i. Soil moisture deficiency reduced susceptibility to ozone,
       PAN and $SO_2$.
    j. Salinity reduced susceptibility to ozone.
    k. High nutrients (nitrogen) generally increased suscep-
       tibility.
    l. Soil oxygen deficiency reduced susceptibility.

## Literature Cited

1. Juhren, M., W. M. Noble and F. W. Went.  Plant Physiol. (1957) 32:576-586.
2. Heck, W. W. and J. A. Dugging.  J. Air Pollut. Contr. Assoc. (1967) 17:112-114.
3. MacDowell, F. D. H.  Can. J. Plant Sci. (1965) 45:1-12.
4. Taylor, O. C., W. M. Dugger, Jr., E. A. Cardiff and E. F. Darley.  Nature. (1961) 192:814-816.
5. Dugger, W. M., Jr., O. C. Taylor, C. R. Thompson and E. Cardiff.  J. Air Pollut. Contr. Assoc. (1963) 13:423-428.
6. Taylor, O. C.  J. Air Pollut Contr. Assoc. (1969) 19(5):347-351.
7. Davis, D. D. Ph.D.  Thesis, The Pa. State University, University Park, PA 93p. (1970).
8. Ting, I. P., and W. M. Dugger, Jr.  J. Air Pollut. Contr. Assoc. (1968) 18:810-813.
9. Heck, W. W.  Ann. Rev. Phytopath (1968) 6:165-188.
10. Heck, W. W. and J. A. Dunning and I. J. Hindawi. J. Air Pollut. Contr. Assoc. (1965) 15:511-515.
11. Dugger, W. M., Jr., O. C. Taylor, E. Cardiff and C. R. Thompson.  Proc. Am. Soc. Hort. Sci. (1962) 81:304-15.
12. Dugger, W. M. and I. P. Ting.  Ann. Rev. Plant Physiol. (1970) 21:215-234.
13. Lee, T. T.  Can. J. Bot. (1965) 43:677-685.
14. Hill, A. C., S. Hill, C. Lamb and T. W. Barrett.  J. Air Pollut. Contr. Assoc. (1974) 24(2):153-157.
15. Cameron, J. W. and O. C. Taylor.  J. Environ. Qual. (1973) 2(3):387-389.
16. Heggestad, H. E., F. R. Burleson, J. T. Middleton, and E. F. Darley.  Inter. J. Air Water Pollut. (1964) 8:1-10.
17. MacDowell, F. D. H.  Can. J. Plant Sci. (1966) 46:349-353.
18. Ten Houten, J. G. "Chemicals, environmental pollution, animals and plants."  Mededlingen Fakultett Landbouwwetenschappen, Gent. (1973) 38:511-612.
19. Middleton, J. T.  J. Air Pollut. Contr. Assoc. (1956-57) 6:7-9.
20. Seidman, G., I. J. Hindawi and W. W. Heck.  J. Air Pollut. Contr. Assoc. (1965) 15:168-170.
21. Stolzy, L. H., O. C. Taylor, J. Letey and T. E. Szuszkiewicz.  Soil Sci. (1961) 91:151-155.
22. Brewer, R. F., F. B. Guillemet and R. K. Creveling. Soil Sci. (1961) 92:298-301.
23. Leone, I. A., E. G. Brennan and R. H. Danes.  J. Air Pollut. Contr. Assoc. (1966) 16:191-196.
24. Menser, H. A. and O. E. Street.  Tobacco Sci. (1962) 6:167-171.
25. Oertli, J. J.  Soil Sci. (1958) 87:249-251.
26. Menser, H. A. and H. E. Heggestad.  Science (1966) 153:424-425.

# Effect of Ozone on Plant Cell Membrane Permeability

IRWIN P. TING, JOHN PERCHOROWICZ, and LANCE EVANS

Department of Biology, University of California, Riverside, Calif. 92502

## General Effects

Ozone, a strong oxidant component of photochemical smog is readily formed under conditions of high radiation and hydrocarbon waste. Oxides of nitrogen, mainly nitrogen dioxide arising from high temperature combustion processes, react with molecular oxygen in the atmosphere in the presence of ultraviolet sunlight to produce nitrogen oxide and ozone. When excess hydrocarbons are present the nitrogen oxide product (NO) will be converted to nitrogen dioxide, resulting in high ozone levels ($\underline{1}$).

$$NO_2 \longrightarrow NO + O$$
$$O + O_2 \longrightarrow O_3$$
$$\overline{\rule{0pt}{0pt}\hspace{8cm}}$$
$$NO_2 + O_2 \longrightarrow NO + O_3$$

The toxic effects of ozone in plant systems have been studied for some time, yet the actual mechanisms of injury are not fully understood. In addition to visible necrosis which appears largely on upper leaf surfaces, many other physiological and biochemical effects have been recorded ($\underline{2}$). One of the first easily measurable effects is a stimulation of respiration. Frequently, however, respiration may not increase without concomitant visible injury. Furthermore, photosynthesis in green leaves as measured by $CO_2$ assimilation, may decrease. It is well known that ozone exposure is accompanied by a dramatic increase in free pool amino acids ($\underline{3}$). Ordin and his co-workers ($\underline{4}$) have clearly shown the effect of ozone on cell wall biosynthesis. In addition, ozone is known to oxidize certain lipid components of the cell ($\underline{5}$), to affect ribosomal RNA ($\underline{6}$) and to alter the fine structure of chloroplasts ($\underline{7}$).

In addition to these effects, there is much evidence to suggest that ozone significantly alters the permeability properties of cell membranes. In fact, Rich ($\underline{8}$) proposed that a primary effect of ozone was the destruction of membrane semi-permeability. As early as 1955, Wedding and Erickson (9) observed an altered phosphate and water permeability in plant cells exposed to ozonated hexene. Dugger and Palmer (10), studying rough lemon leaves exposed to ozone, found a definite alteration in permeability to glucose; after acute exposure to ozone, there was a steady increased uptake of glucose by treated leaves compared to that in controls. The exposed leaves absorbed approximately twice as much glucose as untreated leaves after six days. The increased uptake attributable to ozone exposure was due solely to an increase in glucose-U-$^{14}$C permeability and was not the result of internal increased glucose utilization since there was no difference in $^{14}CO_2$ release from controls and exposed tissues.

In addition to studies with whole cells and tissues, it has also been shown that the permeability of isolated organelles can be altered by ozone exposure. For example, Lee (11) showed that ozone altered the permeability of tobacco mitochondria, and Coulson and Heath (12) reported membrane permeability changes of isolated chloroplasts after ozone exposure.

## Cell Permeability

*Age Effects.* We have found it useful to distinguish between the terms "susceptible" and "sensitive" and thus, have defined the age of maximum injury potential as the susceptible age. Variations in the degree of injury during the susceptible stage are referred to in terms of sensitivity. Susceptible leaves will vary in sensitivity primarily as a function of environmental differences. The age-susceptibility phenomenon to oxidant injury is shown in Figure 1. Young expanding leaves pass through a stage at which they are maximally susceptible to oxidant injury. Neither young nor old leaves are particularly affected by ozone, though only young leaves are susceptible to peroxyacetyl nitrate (PAN) injury (2). Sensitivity of a susceptible leaf can be markedly altered by shifts in environmental conditions such as light, water and nutrition.

Despite the lack of visible necrosis of young and old leaves, it is possible to demonstrate alterations in cell permeability by following metabolite uptake after ozone exposure. If Acala SJ-1 cotton leaves are exposed to ozone and allowed to take up sucrose 24 hr later, there is nearly a doubling of the rate of sucrose uptake by leaves which are maximally susceptible to ozone. However, both young and old leaves which show no visible injury do show about a 50% increase in sucrose uptake (Table I). These data suggest that both young and old tissue are injured to some extent by ozone.

Table I.  Uptake of sucrose-U-$^{14}$C by cotton (Acala SJ-1)
                            leaf discs

| Age of cotyledon (days) | Ratio $O_3$/control* | Visible injury** |
|---|---|---|
| 7 | 1.49 | No |
| 14 | 2.13 | Yes |
| 21 | 1.48 | No |

*Ratio of sucrose uptake by ozone exposed discs to uptake by
control discs.

**Visible injury assessed by observation of necrosis on leaf
surface.  Plants were exposed to 0.5 ppm $O_3$ for 1 hr, 24 hrs
prior to the experiment.


Amino Acid Uptake.  If susceptible leaves are exposed to
ozone and then incubated in leucine-U-$^{14}$C, treated leaves absorb
this amino acid at a faster rate than controls (Fig. 2).  If,
after 4 hr of continuous uptake the $^{14}$C-leucine is followed with
nonlabelled leucine, a significant differential utilization of
the incorporated leucine is not observed (Fig. 2).  This suggests
that the greater uptake is not a consequence of more rapid
utilization and that it is probably attributable to increased
permeability.

Despite a significantly greater uptake of $^{14}$C-leucine by
ozonated tissue, greater incorporation of leucine into protein
does not appear to occur.  When incorporation into protein is
expressed as a percentage of $^{14}$C actually taken up, there is no
difference between the untreated and ozone-treated tissue
(Table II).  These data further support the notion that enhanced
uptake is solely a function of increased transport and not of
increased internal utilization.

Since the kinetics of leucine uptake appear to be exponential, a straight line is obtained when the log of uptake is
plotted against the log of time (Fig. 3).  The constant slope and
the lower y-intercept indicate a higher uptake in ozone-treated
tissue compared with untreated tissue.

Carbohydrate Uptake.  Immediately after ozone exposure,
treated bean leaves are observed to take up less glucose-U-$^{14}$C
than untreated leaves.  After a few hours, however, the ozone-
treated leaves begin to assimilate significantly more glucose.
Similarly, if the nonmetabolizable carbohydrate 2-deoxyglucose is
supplied to ozone-treated leaves, enhanced uptake occurs (Fig. 4).
Immediately following ozone exposure 25% less 2-deoxyglucose is
taken up by ozone-treated tissue, but after 4 hr nearly 50% more

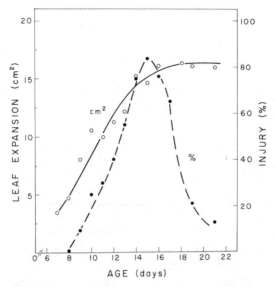

Figure 1. Growth curve for primary leaf of cotton (Acala SJ-1) expressed as cm² expansion as a function of age from seed in days. Also graphed is the visible injury 24 hr after exposure to 0.7 ppm ozone for one hr expressed as % of leaf surface affected.

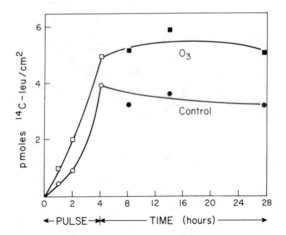

Figure 2. Effect of ozone on uptake and incorporation of $^{14}$C-leucine into protein by cotton cotyledon leaf discs. Plants were exposed to 0.4 ppm $O_3$ for 1 hr, 24 hr prior to experiments. Discs were floated on buffer and incubated in $^{14}$C-leucine for up to 4 hr and were then transferred to excess cold leucine to "chase" the incorporated $^{14}$C-leucine for a subsequent 24 hr period. The data show that ozone-treated tissue incorporated more leucine into protein but do not indicate real differential effects on protein hydrolysis.

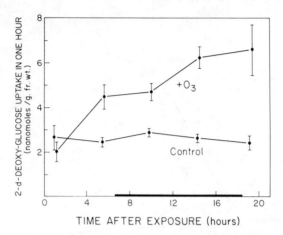

*Figure 3.  Logarithmic plot of $^{14}$C-leucine uptake by cotton leaf discs. Data are from Fig. 2.*

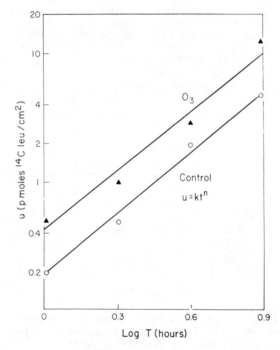

American Journal of Botany

*Figure 4.  Time course of uptake rate of 2-deoxyglucose by bean leaf discs following exposure of intact plants to 0.4 ppm ozone. Two standard deviations about the mean are shown. Data of Perchorowicz and Ting (17).*

uptake has occurred and after 20 hr this reaches a level nearly
300%.

Table II.  Uptake and incorporation into protein of $^{14}$C-leucine
by $O_3$ treated cotton plants

| Treatment | Uptake | | | Turnover | | |
|---|---|---|---|---|---|---|
| | pmoles[a] | cpm[b'] | %[c] | pmoles[a] | cpm[b] | %[c] |
| Control | 5.2 | 12.9 | 19 | 3.6 | 7.3 | 24 |
| Ozone | 5.9 | 14.7 | 18 | 5.7 | 10.7 | 24 |

[a] pmoles $^{14}$C-leucine/cm$_2$ in TCA precipitate.
[b] cpm x $10^{-4}$ $^{14}$C-leucine taken up by leaf discs.
[c] % $^{14}$C-leucine incorporated into protein as a function of
that taken up.

Plants exposed to 0.7 ppm $O_3$ for 1 hr, 48 hrs before experiment.
Uptake    = 4 hr incubation
Turnover = chase experiment 18 hr after uptake period.


As noted with the amino acid uptake data, a logarithmic plot
of deoxyglucose uptake yields a straight line (Fig. 5).  There
is less uptake immediately after exposure of treated tissue, but
after 24 hr a 2- to 3-fold greater uptake is noted.  Subsequent
experiments confirmed that 2-deoxyglucose is not metabolized;
therefore, increased uptake can be attributed to increased permeability and transport rather than to greater internal utilization.
The uptake function appears to be of the form
$$U = Pt^m$$
where P is a function of permeability and m is a function of the
uptake mechanism.  It is possible, therefore, to evaluate the
effects of ozone by a log transformation,
$$\log u = m\log t + \log P$$
and evaluate m and P from a linear graphical plot.  In all cases
m remains unchanged, but P is increased by ozone (Fig. 3 and 5).
As shown in Table III, there is no significant difference in
glucose uptake by cotton leaf discs immediately after exposure,
but after 24 hr the uptake has nearly doubled.  The percentage
distribution of radioactivity in various extraction fractions is
approximately the same in control and ozone-treated tissues.  $CO_2$
evolution from the added glucose is virtually identical in
control and treated tissues.  Isotope incorporation into the water
soluble fraction, which would include amino acids, organic acids,
and soluble carbohydrats, is also quite similar.  Some distribu-
tional differences in the insoluble fraction are noted between

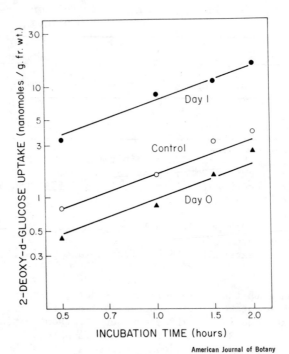

American Journal of Botany

*Figure 5.   Logarithmic plot of 2-deoxyglucose uptake by bean leaf discs immediately following (day 0) and 24 hr (day 1) after exposure to 0.4 ppm ozone for 1 hr. Data from Perchorowicz and Ting (17).*

Table III. d-glucose-$^{14}$C uptake by ozone treated cotyledons and label distribution in extraction fractions

| Hrs after ozone | uptake (nanomoles/g fr wt) | % of d-glucose in Various Extraction Fractions*: | | | |
| --- | --- | --- | --- | --- | --- |
| | | Aqueous | Chloroform | Insoluble | $^{14}CO_2$ |
| Control | 4.7 ± 0.7 | 71 ± 8 | 2.9 ± 1.9 | 15 ± 7 | 12 ± 2 |
| 1 | 5.4 ± 1.1 | 63 ± 10 | 2.5 ± 1.2 | 24 ± 13 | 10 ± 3 |
| 24 | 10.5 ± 5.6 | 77 ± 1.1 | 2.2 ± 1.1 | 10 ± 1 | 11 ± 1 |

Cotton plants were exposed to 1 ppm ozone for one hr and 40 discs were cut from two cotyledons of two plants for uptake of $^{14}$C-UL-d-glucose (4 hours). Each exposed value represents three determinations and each control value represents six determinations.

* ± two standard deviations.

the exposed and control tissues 24 hr after ozone exposure, but
it is difficult to assess the significance of these results.

Nucleotide Uptake. When uracil labelled with tritium was
supplied to 15-day-old ozone-treated cotton leaves, a 2-fold increase in uracil incorporation into the RNA fraction was observed
(Table IV).  In this one particular case, there appeared to be a
greater percentage of uracil incorporation into RNA in treated
tissue than in control tissue.

Table IV.  Incorporation of Uracil - $^3$H into the RNA fraction of
15-day-old $O_3$ treated cotton cotyledon leaves

| Control | Ozone |
|---|---|
| 207,700* | 420,000 |
| 3.1** | 5.5 |

*cpm incorporated into RNA fraction of 40 0.2 cm$^2$ leaf discs.
**% of Uracil incorporated as a fraction of that taken up by
  the discs.

Plants were treated with $O_3$ for 1 hr (0.7 ppm) and supplied with
Uracil for 4 hr, 24 hr after exposure.  RNA was collected on
Millipore filters and counted directly.

Water Transport.  Water transport can conveniently be
summarized by the following expression,
$$J_v = L_p ( \Delta P - \sigma RT\Delta C_s )$$
where $J_v$ is the volume flow of water across the membrane, $L_p$ is
the hydraulic conductivity coefficient and a direct measure of
membrane permeability to water, $\Delta P$ is the pressure differential
across the membrane, $\sigma$ is the reflection coefficient with limits
of 0 and 1 (where 1 indicates complete impermeability with
respect to solutes), R and T are the gas coefficient and absolute
temperature respectively, and $\Delta C_s$ is the solute concentration
difference across the membrane.  Hence, evaluation of $L_p$ will
give an indication of water permeability and evaluation of $\sigma$ will
estimate solute permeability.
The coefficient of hydraulic permeability ($L_p$) of ozone-
treated bean leaf tissue tends to decrease when measured by water
loss or uptake (Fig. 6).  Here, ozone-treated tissue was equilibrated with 0.2 M mannitol (approximately isotonic) immediately
after exposure.  The tissue was then either allowed to take up
tritiated water or, after a period of tritiated water uptake,
allowed to lose tritiated water into a mannitol solution.  In
both the influx and efflux experiments, ozone-treated tissue
transported tritiated water at a lower rate than control tissue.

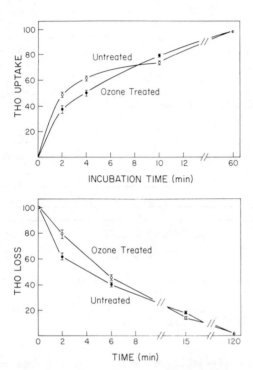

*Figure 6. Time course of tritiated water up-take or release from bean leaf discs previously exposed to 0.5 ppm ozone for 1 hr immediately prior to experiment. Upper: Discs were pre-incubated in 0.2 M mannitol for 2 hr then trans-ferred to tritiated water. Lower: Discs were preincubated in tritiated water then transferred to 0.2 M mannitol. Data of Evans and Ting (13).*

Similar experiments in which tissue was allowed to hydrate or
dehydrate in hypotonic or hypertonic solutions also showed a
lower rate of water transport for ozone-treated tissue (13).
Hence, the coefficient of hydraulic permeability ($L_p$) appears to
be decreased by ozone exposure.

The reflection coefficient ( $\sigma$ ) was estimated by allowing
leaf tissue to equilibrate first in distilled water, then in a
0.4 M mannitol solution for two hr and finally in a distilled
water solution overnight.  The difference in weight between the
initial distilled water equilibration and final distilled water
equilibration was assumed to be an estimate of internal solute
leakage and, therefore, a direct estimate of $\sigma$ .  The data in
Table V shows that loss of solute by the tissue is significant
after ozone fumigation and verifies the predicted decrease in
the reflection coefficient.


Table V.  Effect of ozone on reflection coefficient ($\sigma$)
            of bean leaves*

| Treatment | $\Delta$ w | |
|---|---|---|
| Control | 102.40 $\pm$ 1.72 | 106.26 $\pm$ 0.84 |
| 0.5 ppm $O_3$ (1 hr) | 87.70 $\pm$ 0.87 | 90.50 $\pm$ 1.36 |

*Data of Evans and Ting (1973).
$\Delta$w = fresh wt change before and after 0.4 M mannitol treatment.
    Data for two trials are given.


It is rather difficult to rationalize a decreased membrane
permeability to water ($L_p$) because of oxidant exposure.  We
suspect, therefore, that the apparent decreased water permeability results in fact from a decreased reflection coefficient leading to solute loss and hence an apparent lower water transport
rate.  In any case, these data clearly demonstrate the occurrence
of oxidant-induced alterations in membrane properties.

Salt Transport.  The effects of ozone on membrane permeability can also be assessed by estimating salt leakage from treated
tissue.  In one study, susceptible bean plants were allowed to
take up [86]RbCl for 24 hr prior to ozone exposure.  After exposure, leaf discs were placed in a desorption solution containing 0.5 M $CaSO_4$ and 2 mM KCl and the rate of [86]Rb leakage into
the desorption solution was determined.  The initial loss was
indistinguishable between treated and untreated plants and we
assume that it represented exchange from free space.  Then, for
an extended period, treated tissue exhibited a linear loss of

[86]Rb into the desorption solution, indicating permeability alterations in ozone-treated samples (Fig. 7).

The uptake of potassium by leaf tissue is know to be a function of the energy sources available for transport (14) and significantly more uptake occurs in the light than in the dark. Ozone treatment decreases K uptake under both light and dark conditions. The ozone effect is greater in the light than in the dark (Fig. 8). Ozone may therefore alter the capacity of leaves to take up and accumulate potassium. Though these data once more suggest the occurrence of alterations in membrane properties, the results might also stem from a reduction of the energy sources for transport.

## Conclusions and Summary

The specific cause of membrane alteration by ozone is difficult to assess, though it is undoubtedly related to the oxidizing properties of ozone. It is fairly well established that ozone irreversibly oxidized sulfhydryl groups (2). The oxidant probably also plays a role in lipid peroxidation (16). Regardless of the specific mechanism, a variety of data, including our own, suggests that the initial oxidant effects do indeed involve modifications of membrane properties. These alterations can be measured in plant tissue in which visible symptoms such as necrosis have not appeared. Furthermore, the effects are measureable immediately upon ozone exposure. We conclude that cellular membranes are the initial targets for ozone injury and that many of the subsequent symptoms resulting from ozone exposure, including effects on respiration and photosynthesis, alterations in metabolite concentrations (such as the increase in free pool amino acids), and the formation of crystalline inclusions within chloroplasts, are secondary effects resulting from early changes in membrane properties. Many of these secondary effects may result from dessication following membrane permeability changes (14).

Acknowledgement. The initial portion of this research was supported by Agricultural Research Service, U.S.D.A., contract No. 12-14-100-9493(34) administered by Crops Research Division, Beltsville, Maryland. Appreciation goes to Dr. S. K. Mukerji and Kay Jolley for their participation. The final stages of the work in which John Perchorowicz and Lance Evans participated was supported by Federal Funds from the Environmental Protection Agency under grant number 801311.

Mention of a trademark name or a proprietary product does not constitute a guarantee or warranty of the product by the USDA, and does not imply its approval to the exclusion of other products that may be suitable. The authors gratefully acknowledge the American Journal of Botany and Atmospheric Environment for permission to report previously published material.

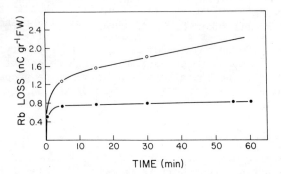

*Figure 7.   Intact plants were watered with a $^{86}$RbCl solution 24 hr prior to exposure to 0.5 ppm ozone for 1 hr.  Discs were then floated on a 0.5 mM CaSO$_4$ plus 2.0 mM KCl solution and $^{86}$Rb leakage was measured.  Closed circles are control plants. Data of Evans and Ting (13).*

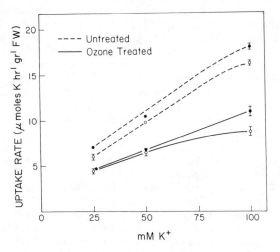

American Journal of Botany

*Figure 8.   Plot of $^{86}$Rb-K uptake by bean leaf discs as a function of K concentration.  Light and dark treatments for control and ozone treated plants are shown. Data of Evans and Ting (15).*

## Literature Cited

1.  Stephens, E. R. J. Air Pollut. Contr. Assoc. (1969) 19: 181-185.
2.  Dugger, W.M. and Ting, I. P. Ann. Rev. Plant Physiol. (1970) 21: 215-234.
3.  Ting, I. P. and Mukerji, S. K. Amer. J. Bot. (1971) 58: 497-504.
4.  Ordin, L. and Hall, M. A. Plant Physiol. (1967) 42: 205-212.
5.  Tomlinson, H. and Rich, S. Phytopathol. (1970) 60:
6.  Chang, C. W. Phytochemistry (1972) 11:  1347-1350.
7.  Thomson, W. W., Dugger, W. M. and Palmer, R. L. Can. J. Bot. (1966) 44:  1677-1682.
8.  Rich, S. Ann. Rev. Phytopathol. (1964) 2: 253-266.
9.  Wedding, R. T. and Erickson, L. C. Amer. J. Bot. (1955) 42: 570-575.
10. Dugger, W. M. and Palmer, R. L. Proc. First Intl. Citrus Symp. Riverside, CA (1969) 2: 711-715.
11. Lee, T. T. Plant Physiol. (1968) 43: 133-139.
12. Coulson, C. and Heath, R. L. Plant Physiol. (1973) 53: 32-38.
13. Evans, L. and Ting, I. P.  Amer. J. Bot. (1973) 60:  155-162.
14. Rains, D. W. Plant Physiol. (1968) 43:  394-400.
15. Evans, L. S. and Ting, I. P.  Atmosp. Environ,  In press.
16. Goldstein, B. D. and Balchum, O. J.  Proc. Soc. Exp. Biol. Med.  (1967) 126: 356-358.
17. Perchorowicz, J. T. and Ting, I. P.  Amer. J. Bot.  (1974) 61: (Aug. Issue)

# Reaction of Ozone with Lysozyme

F. LEH and J. B. MUDD

Department of Biochemistry and Statewide Air Pollution Research Center,
University of California, Riverside, Calif. 92502

Abstract

Hens egg lysozyme is inactivated by ozone in 1.5-2.5 molar
ratios. The pH optimum for the native and ozonized lysozyme is
the same. Ozonized lysozyme behaves differently from native
lysozyme on ion exchange columns and in polyacrylamide gel
electrophoresis but the product behaves as a single polypeptide
chain in both analyses. Amino acid analyses showed that at
95-100% inactivation the modified residues are tryptophan,
tyrosine and methionine. Changing the pH of the reaction mixture
from 4.6 to 10.2 decreases the amount of tryptophan reacting
(1.8 to 1.3 moles/mole lysozyme), increases the amount of
tyrosine reacting (0.13-0.73 moles/mole lysozyme), and does not
change the amount of methionine reacting (0.8 moles/mole
lysozyme). Analysis of the fragments obtained by cyanogen
bromide cleavage shows that methionine 105 is converted to methionine sulfoxide. Reversion of this residue to methionine does
not restore activity. Digestion of the ozonized lysozyme with
trypsin and isolation of the tryptophan-containing fragments
showed that the modified tyrosine is residue 23 and the modified
tryptophan residues are 108 and 111. Circular dichroism spectra
of ozonized lysozyme confirmed the modification of the amino
acids and also showed considerable destruction of three dimensional structure. Experiments with carboxymethyl chitin indicated that the binding site of the ozonized lysozyme is unaffected.

## Introduction

Ozone is an important pollutant of urban atmospheres, and
is generated by photochemical reactions on primary pollutants
emitted from automobile exhaust (1). Ambient ozone concentrations cause acute damage to vegetation. The biochemical basis
for this damage is unknown although several possibilities have
been suggested. This paper presents results applicable to the
study of plant damage even though the subject of study is hens

egg lysozyme.

Exposure of animals to ozone lowers their resistance to bacterial infection of the lungs (2). Ozone exposure also causes changes in the properties of material that can be lavaged from the lungs: there is inactivation of the enzymes acid phosphatase, β-glucuronidase and lysozyme, and there is a tendency for these enzymes to have been released from the alveolar macrophage cells (3). In vivo inactivation of lung lysozyme has been studied (4), and it has also been reported that the content of lysozyme in human tears is lowered during exposure to ambient air pollution (5). These observations have stimulated our interest in a study of the mechanism of lysozyme inactivation by ozone since it may be involved in eliciting the toxic response.

A previous study of the reaction of ozone with lysozyme dissolved in anhydrous formic acid gave rise to the conclusion that the only amino acid residues affected in the early stages of the reaction were the tryptophan residues 108 and 111 (6). Conversion of these residues to N'-formyl-kynurenine did not cause loss in enzyme activity. Imoto et al. (7) have pointed out that this result is anomalous since modifications of tryptophan 108 (e.g. with iodine) normally causes inactivation. We hoped that our study would resolve this anomaly.

## Materials and Methods

Ozonolysis. Ozone was generated from oxygen by silent electric discharge, and bubbled through the reaction mixtures from a capillary tip. The ozone concentration was monitored spectrophotometrically at 350 nm by the KI method as previously described (8).

Material. 3 X crystallized, dialyzed and lyophilized Grade 1 Hen Egg White Lysozyme was obtained from Sigma Chemical Company, and used without further purification. Purity of the protein was verified by both gel electrophoresis on acrylamide and chromatography on a column packed with anion exchange resin (Cl⁻form) Sephadex DEAE. The sample showed a single peak in these analyses.

Chromatographic Analysis. The samples of native and ozonized lysozyme (lysozyme treated with ozone just to the point of complete inactivation) were analyzed by column chromatography. The column (0.8 X 56 cm.) containing DEAE-Sephadex A-50 (Cl⁻form) resin, was equilibrated with 0.1 M Tris Cl buffer, pH 8.3, and loaded with about 2-4 mg of protein. Aliquots eluted with 0.1 M Tris-Cl pH 8.3 were collected and absorbance at 278 nm was measured. The native lysozyme eluted earlier than the ozonized products. This difference may be assoicated with both aggregation of protein and ionic behavior of the residues.

Gel Electrophoresis. This method was used in the determination of the purity of native lysozyme and identification of ozonized products. Different gel concentrations (7,8,9,10%) and buffer solutions (0.25M borate, pH 8.7; 0.025M phosphate, pH 7.1) were tried and the best results were obtained with 7% gel in pH 8.7 buffer.

Enzymic Activity. Lysozyme activity was determined by following the rate of lysis of dried Micrococcus lysodeikticus cells according to the method of Shugar (9). Assays were run at room temperature in 0.1M phosphate buffer pH 7.0, with an enzyme concentration of about 0.05 mg/ml. A solution of native lysozyme at the same protein concentration was always assayed as standard, along with ozonized lysozymes.

Cyanogen Bromide Digestion. In order to cleave the disulfide bonds, the ozonized lysozyme (0.6 μmole) was treated with dithiothreitol (10) (8 mg. 110%) in 9 M urea for 20 hr. The urea was then removed by passing the solution through a column containing Sephadex G-25. Aliquots were collected, checked for absorbance at 278 nm, lyophilized and then allowed to react with cyanogen bromide (11) (50 molar excess relative to methionine) in 70% formic acid (1 ml.) for 24 hr. The mixture was lyophilized and dissolved in 0.2N HAc. Three components separated by passage through a Sephadex G-25 ( 3 X 100 cm.) column were collected for amino acid analysis.

Tryptic Digestion. The ozonized lysozyme solution after treatment with dithiothreitol in 9 M urea at pH 8.0 as described above, was carboxymethylated with sodium iodoacetate (0.02 M) at pH 4.6 and 25°C. for 24 hr, and purified by passage through a Sephadex G-25 column. Hydrolysis of lysozyme by trypsin (1%), and separation of the products were performed under the conditions as previously described by Jolles et al. (12).

Characterization of Enzyme-Substrate Complex by use of CM-Chitin (Carboxymethyl Chitin). CM-chitin was prepared by carboxymethylation of chitin according to the method of Imoto, Hayashi and Funatsu (13). The ozonized lysozyme (1.3 mg) solutions at different pHs were neutralized with NaOH or HCl to pH 8.0 and the poured into the column (1.5 X 4 cm.) containing white cotton-like Cm-chitin (~65 mg.), which was equilibrated with 0.1 M Tris-Cl buffer pH 8.0. Aliquots were eluted first with 0.1 M Tris-Cl pH 8.0 and then with 0.2 M HAc. The absorbance of the fractions was measured spectrophotometrically at 280 nm.

Reduction of Ozonized Lysozyme. The reduction of methionine sulfoxide residues of the photo-oxidized lysozyme was achieved by allowing the enzyme to react with 2-mercaptoethanol in aqueous

solution (14).  The same procedure was applied to reduce the
methionine sulfoxide residue of ozonized lysozyme.

10 mg of ozonized lysozyme in 10 ml of 0.1 M phosphate
buffer at pH 7.0 was allowed to react with 200 ml of 5% aqueous
2-mercaptoethanol at room temperature under nitrogen atmosphere
for 24 hr.  After extensive dialysis against several changes of
cold, distilled water, the pH of the thiol-free solution was adjusted to 8.3 by the addition of 0.1 M Tris-Cl buffer solution.
The solution was kept at 38°C for 12 hr in the presence of a
trace of 2-mercaptoethanol (0.2 mg) and then concentrated, chromatographed on Sephadex G-25 column (1.5 X 100 cm.) and eluted
with 0.2 N acetic acid.  The fraction which had absorption at
278 nm was collected and lyophilized.  The recovered product was
dissolved in 0.1M phosphate buffer for the lytic activity test.

Circular Dichroism Measurements.  Circular dichroism measurements were carried out by using a Cary model 6002 spectro-
polarimeter calibrated with d-10 camphor sulfonic acid.  All
measurements were run at room temperature in the same 1 cm,
quartz cell over the near-ultraviolet region (250-330 nm).
Protein concentrations were approximately 0.4%.  Circular
dichroism data were presented as mean residue ellipticity [$\theta$],
in degrees X $cm^2$ X $decimole^{-1}$.  The same mean residue weight
112.4 was employed for native lysozyme and its ozonized products,
since the deviation caused by the ozonolysis of few amino acid
residues of lysozyme is far less than the experimental error.

Amino Acid Analyses.  Samples of native and ozonized lysozymes were hydrolyzed in evacuated, sealed tubes with 6N hydro-
chloric acid at 110°C for 24 hr.  After being cooled to room
temperature, the solution was adjusted to pH 2 with NaOH solution
and brought to mark in a volumetric flask with sodium citrate
buffer pH 2.3.  A portion containing a suitable amount of the
amino acids was applied to a Beckman 120B amino acid analyzer.

Methionine, methionine sulfoxide and tryptophan were determined after alkaline hydrolysis, since they are known to degrade
during acid hydrolysis.  For this prupose, 3N NaOH was used.
The hydrolyses were performed in sealed evacuated silico tubes
at 100°C for 17 hr.  Then the solutions were cooled, acidified to
about pH 2 with conc. HCl and analyzed for the amino acid composition.

Modification of Tryptophan Residues.  Tryptophan residue
also could be determined quantitatively by a modification with a
sulfenylating agent such as 2-nitrophenyl sulfenyl chloride in
30% acetic acid (15).  Since both oxidized and sulfenylated
tryptophan gave characteristic absorption at 365 nm, the extent
of tryptophan modification was calculated from the mean difference between the sulfenylated protein and the kynurenine.  The
absorption contributed by kynurenine was comparatively weak.

Spectrophotometric Measurements. The measurements of ultra-violet spectra were carried out with a Cary 15 spectrophotometer. The contents of tryptophan and tyrosine were calculated directly from two absorbances at 288 and 280 nm by solving two simul-taneous equations as reported by Edelhoch (16).

## Results

Loss of the Activity and the Change in Primary Structure. Ozone causes the loss of enzymic activity in lysozyme as shown in Figure 1. The enzyme is inactivated linearly until the activity is reduced to 5%. Prolonging exposure of lysozyme to ozone causes further decrease in enzymic activity. Control ex-periments, using gas streams without ozone, were carried out in the same conditions and over the same periods. Figure 2 shows that the activity of both control and ozonized lysozyme varies with pH change. Although the ozonized product shows more ex-tensive dependence on pH than the native one, the optimum is the same.

In order to analyse the properties of the inactivated lyso-zyme, the enzyme was subjected to further analysis by electro-phoresis and chromatography. Both analyses of the samples at various pHs, inactivated to the extent over 95%, indicated that the product is composed of one peak. The ozonized lysozyme moved slower than the native lysozyme on DEAE-Sephadex and the products at different pHs were readily distinguished from each other (Fig. 3). However, a diffuse band was observed for ozonized lysozyme as distinct from a sharp band for native lysozyme in polyacrylamide gel (17). The presence of only one band suggests that the ozonolysis does not cause the cleavage of peptide bonds and the remaining activity is not due to the presence of a small amount of unmodified lysozyme.

Previous study of the reaction of ozone with lysozyme in anhydrous formic acid has demonstrated that investigation of the variations in the ultraviolet spectra of lysozyme provided use-ful information about the alterations of primary structure brought about by ozone (6). The ultraviolet spectrum of lysozyme results principally from four chromophoric amino acid residues: tryptophan, tyrosine, cystine, and phenylalanine with the major contributions from tryptophan ($E_{288}$ = 4815; $E_{280}$ = 5690) and tyrosine ($E_{288}$ = 385; $E_{280}$ = 1280) residues. The conversion of tryptophan into N'-formyl-kynurenine by ozone can be followed by observing a decrease of the maximum absorbancy at 280 nm while two peaks arise at 260 and 320 nm, corresponding to the two maxima of N'-formyl-kynurenine. The more rapid disappearance of absorbance at 280 nm as the pH was changed from 4.6 to 10.2 and the lesser formation of N'-formyl-kynurenine in the more basic solvent, suggested that the oxidation of tyrosine residues occurs and is pH-dependent. Amino acid analyses reveal that cystine and phenylalanine did not react with ozone under these reaction

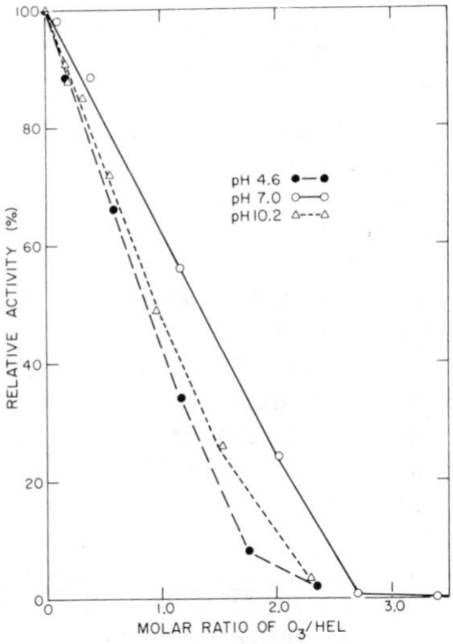

Figure 1. *Inactivation of lysozyme by ozone. Volumes of 4.8 ml of 0.1M buffer containing 2.814 mg of lysozyme were exposed to a gas stream of ozone in oxygen (1.8 nmole/min ozone). Aliquots were removed for enzyme assay as described in Materials and Methods.*

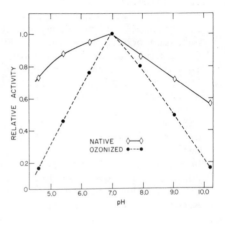

Figure 2. *pH dependence of native and ozonized lysozyme activity. Volumes of 1.0 ml of 0.1M buffer containing 1.91 mg lysozyme were exposed to a gas stream of $O_3/O_2$ at a flow rate of 20 ml/min for 8 min. Ozone delivery was 1.5 nmoles/min. Samples were assayed at the different pHs as described in Materials and Methods.*

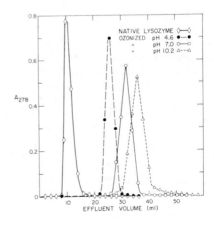

Figure 3. *Chromatography of native and ozonized lysozyme on DEAE Sephadex. Samples of lysozyme at the three different pHs were inactivated 95–100% by ozone. The samples were applied to a column of DEAE-Sephadex A-50 (0.8 × 56 cm) and eluted with 0.01M Tris-HCl pH 8.3 and a gradient of 0.01M NaCl. Fractions of 2 ml were collected. The results of four columns are plotted.*

conditions; this fact and the observation of u.v. spectra that
kynurenine makes only a small contribution ($E_{280}$ = 210) to the
total absorbance at 280 nm make it reasonable to calculate the
extent of tryptophan and tyrosine modification from absorbance
measurements at 288 and 280 nm by the method of Edelhoch (16).
Further confirmation can be obtained by a determination of
tryptophan from the absorption band at 365 nm developed by
sulfenylation with 2-nitrophenyl sulfenyl chloride (15). Both
these methods give results in good agreement with those obtained
from amino acid analyses (Table I).

At pH 10.2, the ultraviolet spectrum of lysozyme (Fig. 4)
shows a sharp change of 280 and 288 nm bands after treatment with
equimolar amounts of ozone. At this point, the lysozyme retained
40% of its activity. This behavior contrasted with the linear
changes of 280 and 288 nm absorbance at pH 4.6 and 7.0.

Location of the Modified Residues. In order to determine
the location and number of amino acid residues reacted with
ozone, the ozonized samples were allowed to undergo cyanogen
bromide cleavage and trypsin digestions.

Bonavida, Miller and Sercarz (11) have demonstrated that the
reaction of native lysozyme with CNBr is solvent-dependent. No
reaction occurs in either 2N HCl or 10% formic acid, but the
methionine can be quantitatively converted into homoserine
lactone with concomitant cleavage of the methionyl peptide in
70% formic acid. Cleavage of the methionyl peptide bonds of HEL
should result in the formation of two new free amino groups at
residue 13 (lysine) and residue 106 (asparagine); the reaction of
CNBr with ozonized HEL would be expected to give partial cleavage
of the methionyl peptide bond, because the oxidized product of
methionine---methionine sulfoxide will not react with cyanogen
bromide (18) (Scheme 1). For separation of CNBr-treated HEL
peptides, the ozonized HEL was pretreated with dithiothreitol
(DTT) in the presence of 9M urea to cleave the disulfide bonds.
After removal of CNBr and formic acid under reduced pressure, the
product was dissolved in 0.2N HAc and chromatographed on a
Sephadex G-25 column. Three peaks were obtained as shown in
Figure 5. Amino acid analyses (Table II) indicated that the
first of these peaks is the sum of the core polypeptide $L_{11}$
(residues 13-105), plus $L_{1V}$ (residues 13-129); the second is the
carboxylterminal peptide $L_{111}$ (residues 106-129); and the third
is the amino-terminal dodecapeptide $L_1$ (residues 1-12). $L_{11}$,
$L_{111}$ and $L_{1V}$ absorb at 280 nm, whereas $L_1$ has no tryptophan or
tyrosine residues and shows no absorption at 280 nm, but can be
detected with ninhydrin after alkaline hydrolyses. This result
reveals that ozone causes the oxidation of the methionine residue
105 of lysozyme.

In order to determine the positions of the modified tryptophan residues, the ozonized lysozyme at pH 7.0 was reduced, carboxymethylated and digested with trypsin. The tryptophan-

Table I.  Amino acid content of untreated and ozonized lysozyme

| Amino acid | Untreated (unit) | reacted with ozone | | | | | | | | | | |
| | | pH 4.6 | | | pH 7.0 | | | pH 10.2 | | | |
| | | A* | B* | C* | A | B | C | A | B | C |
|------------|------------------|------|------|------|------|------|------|------|------|------|
| tryptophan | 6.0 | 1.83 | 1.81 | 1.84 | 1.63 | 1.62 | 1.65 | 1.34 | 1.31 | 1.36 |
| tyrosine | 3.2 | 0.15 | 0.18 | 0.13 | 0.45 | 0.47 | 0.44 | 0.71 | 0.73 | 0.72 |
| methionine | 1.90 | 0.84 | -- | -- | 0.77 | -- | -- | 0.72 | -- | -- |

A* = amino acid analyses; B* = Edelhoch's method (16); C* = Scoffone's method (15).

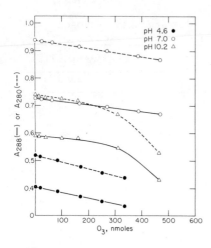

*Figure 4. Changes in UV absorbance of ozonized lysozyme. Volumes of 3.0 ml of 0.1M buffer containing 0.59–1.11 mg lysozyme (initial amounts of protein were different) were exposed to $O_3/O_2$ at 20 ml/min. Ozone delivery was 1.6 nmoles/min. Absorbance spectra were recorded at intervals and changes at 280 and 288 nm were plotted.*

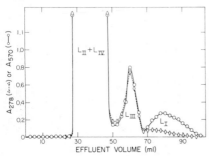

*Figure 5. Separation of ozonized lysozyme CNBr fragments. The ozonized lysozyme was treated with CNBr as described in the Materials and Methods. The peptide mixture was applied to a column of Sephadex G-25 (3 × 100 cm) and eluted with 0.2N HAc at a flow rate of 0.5 ml/min. Fractions were assayed by measuring UV absorbance on ninhydrin color after alkaline hydrolysis.*

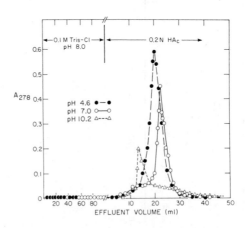

*Figure 6. Binding of ozonized lysozyme to CM-chitin. Samples of lysozyme (1.3 mg) ozonized at different pHs to > 95% inactivation were applied to a column of CM-chitin (1.5 × 4 cm). The elution sequence was firstly Tris-Cl and secondly 0.2N HAc.*

Table II.  Amino acid composition of ozonized and cyanogen
bromide treated Hen Egg-White Lysozyme

| Amino acid | $L_1$(100%) | | $L_{111}$(23%) | | $L_{11}$(23%)+$L_{1V}$(77%) | |
|---|---|---|---|---|---|---|
| | theory | found | theory | found | theory | found |
| Lysine | 1.0 | 1.13 | 0.23 | 0.26 | 4.77 | 4.72 |
| Histidine | | | | | 1 | 0.99 |
| Arginine | 1.0 | 0.88 | 0.92 | 1.17 | 9.08 | 9.04 |
| Aspartic acid | | | 0.69 | 0.58 | 20.31 | 19.85 |
| Threonine | | | 0.23 | 0.24 | 6.77 | 6.93 |
| Serine | | | | | 10.00 | 9.86 |
| Glutamic acid | 1.0 | 0.84 | 0.23 | 0.29 | 3.77 | 3.82 |
| Proline | | | | | 2.00 | 1.93 |
| Glycine | 1.0 | 0.91 | 0.46 | 0.43 | 10.54 | 10.71 |
| Alanine | 3.0 | 2.95 | 0.69 | 0.74 | 8.31 | 8.24 |
| Half-cystine | 1.0 | 0.91 | 0.46 | 0.58 | 6.54 | 6.40 |
| Valine | 1.0 | 0.93 | 0.46 | 0.54 | 4.54 | 4.52 |
| Methionine* | | | | | | |
| Isoleucine | | | 0.23 | 0.29 | 5.77 | 5.73 |
| Leucine | 1.0 | 0.96 | 0.23 | 0.41 | 6.77 | 6.80 |
| Tyrosine* | | | | | 3.00 | 2.55 |
| Phenylalanine | 1.0 | 0.86 | | | 2.00 | 2.21 |
| Tryptophan* | | | 0.69 | 0.31 | 5.31 | 4.06 |

*ozonized residues.

containing peptides were separated by the method described by
Jolles et al. (12). The amino acid analyses data are listed in
Table III. The loss of tryptophan with the appearance of N'-
formyl-kynurenine in peptide $T_{16}$ (residues 98-112), and the
complete resistance of tryptophan in peptides $T_{10}$ (residues 117-
125), $T_{14}$ (residues 22-23), and $T_{17}$ (residues 62-68), indicated
that of the six tryptophan residues only 108 and 111 are oxidized. The change of the content of tyrosine residue in peptide
$T_{14}$ shows that only the tyrosine at position 23 is modified.

  Reduction of Ozonized Lysozyme. Methionine sulfoxide can
revert to methionine with the generation of the lytic activity
for photo-oxidized lysozyme (14). We tested whether the ozonized
lysozyme could be reactivated by chemical reduction. The ozonized lysozyme was treated with 2-mercaptoethanol, dialysed,
purified by passage through a column of Sephadex G-25 and lyophilized. The product showed no increase in its lytic activity.
This is not surprising because residues other than methionine are
oxidized, but it may be concluded that the oxidation of methionine alone cannot account for enzyme inactivation.

  Estimation of the Binding Site. Tryptophan-108 shows a
specific reaction with iodine, distinguishing it from other tryptophan residues of lysozyme. When try-108 is selectively oxidized by iodine, lysozyme completely loses its activity. Never-
theless, the lysozyme still shows the ability to form an enzyme-
substrate complex with CM-chitin. This observation contributes
to the conclusion that try-62 is an essential binding site for
a complex formation (13). All ozonized lysozymes formed strong
complexes with CM-chitin and could only be eluted by 0.2N $HA_C$
(Fig. 6). This further confirms that two tryptophan residues
(108 and 111) are indispensible for the hydrolytic action of
lysozyme, and that inactivation by ozone cannot be attributed to
inhibition of substrate binding capability.

  Conformation. The near ultraviolet CD (circular dichroism)
spectra of native lysozyme at pH 4.6, 7.0 and 10.2 (Fig. 7A) are
characterized by three positive peaks at 294, 288 and 282 nm and
a large negative band at 268-266 nm. These circular dichroism
patterns are much like those reported by others (19,20,21).
Increasing pH brings about increasingly positive ellipticity
values at 255 nm and at 295 nm, as well as alternation in the
positions of the large negative band at 268 nm. The change in
ellipticity at 255 nm parallels the ionization of the tyrosine
residues in which two of the three tyrosine residues are reversibly ionized. This interpretation has been reached by the
tyrosine titration (20,22).
  Figure 7B represents the near-ultraviolet CD spectra of
ozonized lysozyme in three different buffer solutions. Comparing
the CD spectra of native and ozonized lysozymes, one observes

Table III.  Trypsin digestion fractions of ozonized Hen Egg White Lysozyme

No. of tryptic peptides according to Jolles et al. (12).

| Amino acid | T10 theory | T10 found | T14 theory | T14 found | T16 theory | T16 found | T17 theory | T17 found |
|---|---|---|---|---|---|---|---|---|
| Glycine | 1 | 1.04 | 2 | 2.06 | 2 | 2.03 | 1 | 1.03 |
| Alanine | 1 | 1.12 | 2 | 2.15 | 2 | 2.04 | | |
| Serine | | | 1 | 0.94 | 1 | 1.01 | | |
| Half-cystine | | | 1 | 0.89 | | | 1 | 0.95 |
| Threonine | 1 | 0.95 | | | | | | |
| Valine | 1 | 0.93 | 1 | 1.16 | 2 | 2.20 | | |
| Leucine | | | 1 | 1.11 | | | | |
| Isoleucine | 1 | 1.02 | | | 1 | 1.15 | | |
| Tyrosine* | | | 1 | 0.55 | | | | |
| Tryptophan* | 1 | 0.97 | 1 | 1.14 | 2 | 0.37 | 2 | 1.93 |
| Aspartic acid | 1 | 1.24 | 1 | 1.16 | 3 | 3.08 | 2 | 2.01 |
| Glutamic acid | 1 | 0.98 | | | | | | |
| Lysine | | | 1 | 0.96 | | | | |
| Arginine | 1 | 1.16 | | | 1 | 0.93 | 1 | 0.88 |

*ozonized residues.

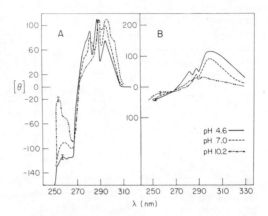

*Figure 7.   CD spectra of native (A) and ozonized (B) lysozyme. Reproducibility of θ for each spectrum was within 5% indicated by vertical bar. Other details are given in Materials and Methods.*

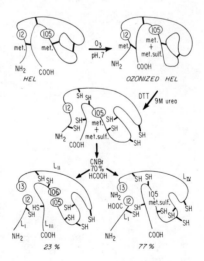

*Scheme 1.   Ozone treatment and CNBr fragmentation of lysozyme.*

several interesting features. In general, the circular dichroism
spectra lose their fine structure; the positive band at 295 nm
and the negative band at 268 nm disappear. At the lower pH, a
new broad peak centered at 300 nm is observed while the inten-
sities of the bands of native lysozyme at 282 nm and 288 nm are
gradually reduced depending on the pH of the solution. A pronounced effect is observed for lysozyme ozonized at pH 10.2
arising from dramatic structural change. As it is known that the
near-ultraviolet CD of lysozyme is entirely dependent on the
native tertiary structure of the protein, the spectrum can be
completely abolished due to unfolding the molecule. This result
closely resembles the effect of guanidine hydrochloride in denaturing lysozyme (23).

## Discussion

Distinct changes in several properties of lysozyme occur
after reaction with ozone. The lytic activity of the ozonized
lysozyme shows the same trend at various pHs as the native enzyme
(Fig. 2); this may suggest that the pK values of the ionizable
groups involved in catalysis have not been altered by ozonolysis.
The amino acid composition of ozonized lysozyme differs from that
of the native enzyme in three residues -- methionine, tryptophan
and tyrosine. None of the other amino acids is affected by ozone.
The extensive loss of enzymic activity must be ascribed to the
oxidative modification of these three amino acid residues in the
lysozyme.

In an earlier experiment, Jori et al. (14) reported that
methionyl residues are important in maintaining the tertiary
structure of lysozyme. The introduction of a polar center into
the aliphatic side chain of methionine, as a consequence of the
conversion of the thioether function to the sulfoxide, may bring
about a structural change of the lysozyme molecule which, in
turn, reduces the catalytic efficiency. When ozonized lysozyme
was treated with 2-mercaptoethanol in an aqueous solution according to the procedure of Jori et al. (14), the enzyme did not show
any increase in its activity. This may be explained in two ways.
In one, such reactions are complicated by many side reactions,
e.g. sulfhydryl-disulfide interchange, aggregation and precipita-
tion of the modified enzyme (24-26). In the other, the failure
to recover the activity of the enzyme may by associated with the
extensive oxidation of other residues.

Tryptophan 108 is recognized to be an active site in promot-
ing the hydrolysis of $\beta(1,4)$-glycosidic linkages between amino
sugar residues in polysaccharide components of the bacterial cell
walls. This residue is shown to occupy the cleft as well as
tryptophan 62 and 63, and is in a hydrophobic region. Tryptophan
residues 62 and 108 are indispensable for the action of lysozyme,
and tryptophan 62 is known to be the only binding site for the
complex formation (13). Oxidation of tryptophan-108 is expected

to cause the lysozyme to completely lose its catalytic activity, but to retain completely its ability to form an enzyme-substrate complex with CM-chitin. This prediction is in agreement with our observation with ozone and others in the reaction of the lysozyme with iodine (27,28). The high susceptibility of tryptophan 108 to electrophiles like ozone and iodine may support the conclusion that the same mechanism operates in both cases. The indole N of try-108 is hydrogen-bonded to a main chain carboxy and the 2-position is in contact with the carboxy of glutamic 35 (27). Both interactions would increase the electron density of the ring and so activate it with respect to attack by the electrophile.

Circular dichroism was used to elucidate the conformational change of lysozyme upon ozonolysis. In native lysozyme, the shorter wavelength negative band is assigned to a disulfide transition and the longer wavelength positive bands are contributed to by tyrosine and tryptophan (29). By systematic examination of the effect of pH and chemical modification, it is possible to correlate the changes of the observed ellipticity from different chromophores with the reaction of ozone. The results in Figure 7A indicate that little alternation in structure occurs by varying pH between 4.6 and 10.2. The oxidation of tryptophan 108 and 111, however, abolishes the positive ellipticity at 294 nm (21), due to the interruption of the coupling of the transition in residues 108 and 111 with identical transitions in other tryptophan residues such as 28, 62, and 63 (30). Other evidence supports this conclusion from the circular dichroism studies on the oxidation of tryptophan 108 by iodine (21) and on the binding of oligomers of N-acetyl-D-glucosamine to try-108 (21). Comparison of the different spectra at three pHs shows that accessible tyrosine residues contribute a positive band near 282 nm (20) and a negative band at 255 nm. Ionization of these tyrosine residues leads to a diminution of both positive and negative ellipticity. The change parallels the change of pH. At high pH, the ionized tyrosine residue is preferentially attacked by the ozone. This independent study further indicates that the pH effects on the reactions of ozone with protein are similar to those with free amino acids (8). Paralleling the appearance of the absorption at 320 nm in u.v. spectra, the new broader band at 299-304 nm in the CD spectrum is tentatively assigned to the contribution from N-formyl-kynurenine.

Straight-chain or large-ring disulfides characteristically generate a CD band between 249 and 260 nm (29,31). The optical activity arises from inherent dissymmetry, from asymmetric perturbation or from both (32). If the externally perturbing group is unaltered relative to a fixed coordinate system, the rotation around the disulfide bond (change in screw configuration) would change the sign of the optically active band. Barnes, Warren and Gordon (19) recently demonstrated that the 258 nm band in lysozyme arises primarily from the intramolecular disulfide bands and alters with the conformational changes. Based on the amino acid analysis showing that the disulfide bonds are not reacted with ozone, the intensity of the CD band near 255-265 nm could be utilized for the diagnosis of structural change. This band is completely abolished in the spectra of ozonized lysozyme (Fig. 7B) indicating the change of structure accompanying the modification of amino acid residues by ozone. Holladay and Sophianopoulos (23) concluded that the bands at 288 and 294 nm are due "nearly completely to the tertiary and quaternary structure" of lysozyme and they are completely abolished during denaturing. Since these bands arise from the chromophores tyrosine and tryptophan which react with ozone, the information from the CD spectra concerning three dimensional structure is less dependable.

## Literature cited

1.  Leighton, P. A. "Photochemistry of Air Pollution," p. 300, Academic Press, New York. (1961).
2.  Goldstein, E., Tyler, W. S., Hoeprich, P. D. and Eagle, C. Arch. Int. Med. (1971) 127:1099-1102.
3.  Hurst, D. J., Gardner, D. E. and Coffin, D. L. Reticuloendothel. Soc. (1970) 8:288-300.
4.  Holzman, R. S., Gardner, D. E. and Coffin, D. L. J. Bact. (1968) 96:1962-1966.
5.  Saspe, A. T., Bonavida, B., Stone, W. and Sercarz, E. E. Amer. J. Ophthal. (1968) 66:76-80.
6.  Previero, A., Coletti-Previero, M.-A. and Jolles, P. J. Mol. Biol. (1967) 24:261-268.
7.  Imoto, T., Johnson, L. N., North, A. C. T., Phillips, D. C. and Rupley, J. A. In "The Enzymes VII." 665-808, ed. Boyer, P. D. Academic Press, New York. (1972).
8.  Mudd, J. B., Leavitt, R., Ongun, A. and McManus, T. T. Atmos. Environ, (1969) 3:669-82.
9.  Shugar, D. Biochim. Biophys. Acta (1952) 8:302.
10. Cleland, W.W. Biochem.(1964) 3:480.
11. Bonavida, B., Miller, A. and Sercarz, E. E. Biochem. (1969) 8:968-79.
12. Jolles, J., Jauregui-Adell, J., Berrier, J. and Jolles, P. Biochim. Biophys. Acta (1963) 78:668-89.
13. Imoto, T., Hayashi, K. and Funatsu, M. J. Biochem. (1968) 64:387-92.
14. Jori, G., Galiazzo, G., Marzotto, A. and Scoffone, E. J. Biol. Chem. (1968) 243:4272-78.
15. Scoffone, E., Fontana, A. and Rocchi, R. Biochem. (1968) 7:971-979.
16. Edelhoch, H. Biochem. (1967) 6:1948-54.
17. Finlayson, A. J. Can. J. Biochem. (1969) 47:31-7.
18. Gross, E. and Witkop, B. J. Biol. Chem. (1962) 237:1856.
19. Barnes, K. P., Warren, J. R. and Gardon, J. A. J. Biol. Chem. (1972) 247:1708-12.
20. Halper, J. P., Latovitzki, N., Bernstein, H. and Beychok, S. Proc. Nat. Acad. Sci. USA (1971) 68:517-22.
21. Teichberg, V. J., Kay, C.M. and Sharon, N. Europ. J. Biochem. (1970) 16:55-59.
22. Ikeda, K. and Hamaguchi, K. J. Biochem. (Tokyo) (1969)66: 513-520.
23. Holladay, L. A. and Sophianopoulos, A. J., J. Biol. Chem. (1972) 247:1976-79.
24. Epstein, C. J. and Goldberger, R. F. J. Biol. Chem. (1963) 238:1380-1383.
25. Kanarek, L., Bradshaw, R. A. and Hill, R. L. J. Biol. Chem. (1967) 240:pc 2755-2757 (1965); 242:3789-3798 (1967).
26. Imai, K., Takagi, T. and Isemura, T. J. Biochem. (Tokyo) (1963) 53:1-6.

27. Hartdegen, F. J. and Rupley, J. A. J. Amer. Chem. Soc. (1967) 89:1743-1745.
28. Blake, C. C. F. Proc. Roy. Soc. (1967) 167 B:435-438.
29. Beychok, S. Proc. Nat. Acad. Sci USA (1965) 53:999-1005.
30. Zuclich, J. J. Chem. Phys. (1970) 52:3586-3591.
31. Beychok, S. and Breslow, E. J. Biol. Chem (1968) 243: 151-54.
32. Linderberg, J. and Michl, J. J. Amer. Chem. Soc. (1970) 92:2619-25.

# 4

# Ozone Induced Alterations in the Metabolite Pools and Enzyme Activities of Plants

DAVID T. TINGEY

Environmental Protection Agency, 200 S.W. 35th Street, Corvallis, Ore. 97330

## Abstract

Acute or chronic ozone exposure may reduce plant growth and cause greater reductions in root growth than in top growth. These growth reductions are associated with metabolic alterations.

When soybean leaves and pine needles were exposed to ozone, there was an initial decrease in the levels of soluble sugars followed by a subsequent increase. Ozone exposure also caused a decrease in the activity of the glycolytic pathway and the decrease in the activity was reflected in a lowered rate of nitrate reduction. Amino acids and protein also accumulated in soybean leaves following exposure. Ozone increased the activities of enzymes involved in phenol metabolism (phenylalanine ammonia lyase and polyphenoloxidase). There was also an increase in the levels of total phenols. Leachates from fescue leaves exposed to ozone inhibited nodulation.

Ozone-induced reductions in root growth resulted from altered foliage metabolism rather than from direct action upon the roots themselves. More specifically the decrease in root growth probably resulted from a reduction in either translocation and/or the quality of the photosynthate translocated to the roots. Soluble carbohydrates and starch were at lower levels in roots of ponderosa pine exposed to low levels of ozone at the end of the growing season, while the levels of amino acids and Kjeldahl nitrogen were higher. Nodulation of legumes exposed to ozone was reduced suggesting a reduction in the amount of nitrogen fixed per plant. Root exudates from plants exposed to ozone also inhibited root growth and nodulation in other plants.

## Introduction

Plant growth and development is a coordinated set of inter-related events. The growth of various plant parts can influence the growth of other parts through metabolic activities (<u>1</u>).

Leaves supply the bulk of the photosynthate that supports plant
growth while roots absorb necessary water and mineral nutrients.
Leaves and roots also synthesize hormones that influence plant
growth and development.  It is apparent that reduced growth of
either root or shoot would be reflected in the growth and metabolite pools of other plant parts.  An air pollutant, such as
ozone, would be expected to reduce plant growth (a) if it impaired a rate limiting step in growth; (b) if it made some plant
system rate-limiting; (c) if it reduced the availability of a
needed metabolite or hormone at the growth site; or (d) if it
caused the formation of a phytotoxic compound.

The object of this paper is to show that both acute and
chronic ozone exposures can exert similar effects on plants by:
(a) reducing foliage and root growth; (b) directly altering
metabolite pools and enzyme activities in foliage; (c) indirectly
altering metabolite pools in roots.  The paper also suggests that
these alterations in metabolite pools in leaves and roots could
be associated with observed reductions in plant growth.

## Effects on Plant Growth

Plant growth may be reduced by single, multiple or chronic
ozone exposures.  Radishes receiving a single ozone exposure at
different stages of plant development exhibited growth reductions
ranging from 2 to 15% for foliage and 15 to 37% for root growth
(2) (Table I).   When radishes received multiple ozone exposures,
growth reductions ranged from 10 to 27% for foliage and from 43
to 75% for root growth (Table I).

Table I.  Reduction in Radish Growth from Single or Multiple
Ozone Exposures[1]

| Time of exposure (days from seeding) | Percent reduction[2] foliage dry wt | Percent reduction[2] root dry wt |
|---|---|---|
| 21 | 2 | 15 |
| 7 | 15 | 23 |
| 14 | 10 | 37 |
| 7 + 21 | 16 | 43 |
| 14 + 21 | 10 | 54 |
| 7 + 14 | 24 | 63 |
| 7 + 14 + 21 | 27 | 75 |

[1] Plants were grown and exposed to 785 $\mu g/m^3$ ozone for 1.5 hr in
a controlled environment facility and harvested at 28 days from
seeding.  Means are based on 24 observations.

[2] Data from "Proceedings Third International Clean Air Congress"
(2) and unpublished data from Tingey and Dunning.

Growth reductions from the multiple exposures were additive to
the effects of the single exposures. For all exposures the reduction in root growth was substantially larger than that in
foliage growth. Chronic ozone exposures reduced the growth of
several plants (Table II). Growth reductions ranged from 5 to
70% for top growth and 16 to 73% for root growth. Root growth
was reduced more than top growth in all treatments except the
tobacco receiving the 200 $\mu g/m^3$ ozone treatment where the reductions were similar in magnitude.

## Changes in Plant Metabolites

In general, the effects of an acute ozone exposure on plant
metabolism will be illustrated using data from soybean, cv. Dare.
The soybeans were exposed to ozone for two hr when the first
trifoliate leaf was 50 to 60% expanded and analyzed for various
metabolites and enzyme activities at 0, 24, 48 and 72 hr following termination of exposure. To illustrate the effects of
chronic ozone exposures, data from Ponderosa pine seedlings will
be used. Ponderosa pine were grown from seed under field conditions and exposed to 200 $\mu g/m^3$ of ozone 6 hr per day throughout
the growing season. Harvests were made at monthly intervals
after the initiation of exposure.

## Changes in Metabolite Pools in Foliage

In soybeans a single acute ozone dose (0, 490 or 980 $\mu g/m^3$
ozone for 2 hr) significantly decreased the level of reducing
sugars immediately following exposure, and the concentrations
remained below the control level for 24 hr (Fig. 1). Levels of
reducing sugars were not different among treatments at 48 hr;
however, after 72 hr the reducing sugar level in plants receiving 980 $\mu g/m^3$ was higher than the control level (3). The ozone
treatment had no effect on starch content.
Soluble carbohydrate levels of seedling Ponderosa pine were
altered by a chronic ozone exposure (Tingey and Wilhour, unpublished). The level of soluble carbohydrates was 10% below the
control level one month after initiation of the exposure
(Fig. 2). At the second and subsequent harvests, there were
higher levels of soluble carbohydrates in the tops of exposed
plants than in the controls. Starch levels in exposed foliage
were elevated above the control after two months of exposure and
stayed there for the remainder of the season.
The ozone-induced depression in soluble sugar levels observed in soybean and pine could have resulted from a depression
in the photosynthetic rate. Hill and Littlefield (4) reported
that 785 to 1175 $\mu g/m^3$ ozone for 1/2 to 1 1/2 hr reduced photosynthesis about 50% in several plant species. Ozone has also
been shown to decrease the photosynthesis of Ponderosa pine (5).

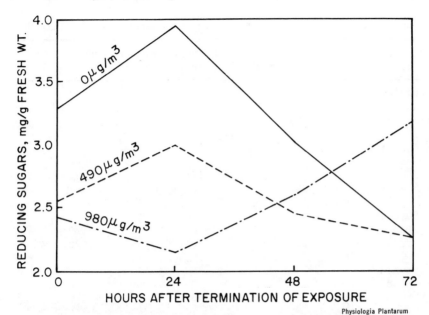

Figure 1.   *Effects of a single ozone exposure on the levels reducing sugars in soybean leaves (3). Each mean is based on four observations.*

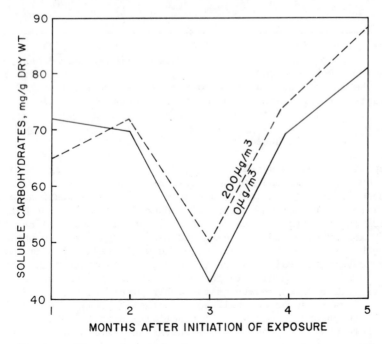

Figure 2.   *Effect of chronic ozone exposure on the soluble carbohydrate levels of seedling Ponderosa pine tops. Each mean is based on nine observations.*

Table II.   Effects of Chronic Ozone Exposures on Plant Growth

| Plant[1] | Ozone conc. ($\mu g/m^3$) | Percent reduction top dry wt | Percent reduction root dry wt |
|---|---|---|---|
| Radish Cherry Belle | 0 | 0 | 0 |
|  | 100 | 10 | 50 |
| Alfalfa Vernal | 0 | 0 | 0 |
|  | 100 | 13 | 22 |
|  | 200 | 36 | 57 |
| Soybean Dare | 0 | 0 | 0 |
|  | 200 | 6 | 28 |
| Soybean Hood | 0 | 0 | 0 |
|  | 200 | 5 | 16 |
| Tobacco Bel W3 | 0 | 0 | 0 |
|  | 100 | 10 | 41 |
|  | 200 | 70 | 73 |

[1]Data for radish is from J. Amer. Soc. Hort. Sci. (23); data for
the other plants is unpublished data of Tingey and Reinert.
Radishes were exposed for 5 weeks, alfalfa was exposed for 3
months and soybean and tobacco were exposed 4 and 5 weeks.   All
plants were grown and exposed under greenhouse conditions.


The ozone-induced increase in the levels of soluble carbohydrates in soybean and Ponderosa pine foliage has also been
reported by other workers (5, 6, 7).   The increase in soluble
carbohydrates in the foliage could have resulted from (a) reduced sugar utilization; (b) reduced sugar translocation, and/or
(c) the fact that the sugar was rendered unavilable for metabolism in products such as glycosides.   In any case, the retention
of carbohydrates in foliage could cause a reduction in photosynthesis by feedback inhibition and would reduce the amount of
photosynthate available for translocation to other plant organs.
The decrease in available photosynthate for translocation could
result in reduced growth of the other plant organs.
     Sugars can be metabolized by either glycolysis or by the
pentose phosphate pathway.   The activities in these pathways can
be determined by measuring the enzyme activities of selected
dehydrogenases in each path.   In soybean, ozone can alter the

activities of selected enzymes in the glycolytic and pentose
phosphate pathways (8). A single ozone exposure (980 µg/m$^3$ ozone
for 2 hr) depressed the activity of glyceraldehyde 3-phosphate
dehydrogenase (GPD) 20% and the activity of the enzyme remained
significantly below the control level for at least 72 hr
(Fig. 3A). Conversely, the activity of glucose 6-phosphate dehydrogenase (G6PD) was unaffected immediately following the ozone
exposure; but within 24 hr, the activity of G6PD was significantly above the control level and it remained there for at least
72 hr (Fig. 3B). Another enzyme in the pentose phosphate pathway, 6-phosphogluconate dehydrogenase, showed similar trends to
G6PD (Tingey, unpublished).

The ozone treatment apparently initiated changes in the
pathways of carbohydrate metabolism, with glycolysis being reduced while the activity of the pentose phosphate pathway was
increased. Rat lung tissue exposed to ozone also exhibited the
ozone-induced depression of GPD and enhancement of G6PD activity
(9, 10). The activation of the pentose phosphate pathway is a
characteristic feature of diseased plants (11, 12).

In plant leaves, nitrate reduction requires NADH produced
by glycolytic activity (13). The ozone-induced depression of
glycolytic metabolism in soybean leaves was also reflected in a
depressed rate of nitrate reduction (3). A single ozone exposure depressed the in vivo nitrate reductase (NR) activity
about 60% (Table III). To determine if ozone affected the NR
protein directly, the in vitro NR activity was determined in
leaf extracts from plants exposed to 0 and 980 µg/m$^3$ ozone.
The ozone treatment had no significant effect on the in vitro
NR activity, indicating that it did not inactivate the NR protein
(Table III). Leaf extracts that would couple the oxidation of
fructose-1, 6-diphosphate to nitrate reduction were prepared from
leaves exposed to either 0 or 980 µg/m$^3$ ozone (3). Ozone depressed the in vitro coupled NR activity 58% (Table III), indicating that the observed ozone depression of nitrate reduction in
the in vivo leaf disk assay resulted from a depression in the
rate of NADH formation by GPD.

In soybean leaves exposed to a single acute ozone dose
(0, 490 or 980 µg/m$^3$ ozone for 2 hr) the amino acid level was
depressed immediately following exposure (Fig. 4A) (3). This
depression of amino acids was similar to that observed in levels
of reducing sugars immediately following exposure and probably
resulted from a depression in photosynthesis. Within 24 hr after
exposure, the amino acid level in plants exposed to 490 µg/m$^3$
had returned to the control level. However, in plants exposed
to 980 µg/m$^3$, the amino acid level increased above that of the
control. Concurrent with the rise in free amino acids was an
increase in protein level in exposed foliage (Fig. 4B). The
increase in amino acids following an ozone exposure has been reported by other workers (6, 14, 15); however, in these cases, the
rise in free amino acid level was associated with a decline in

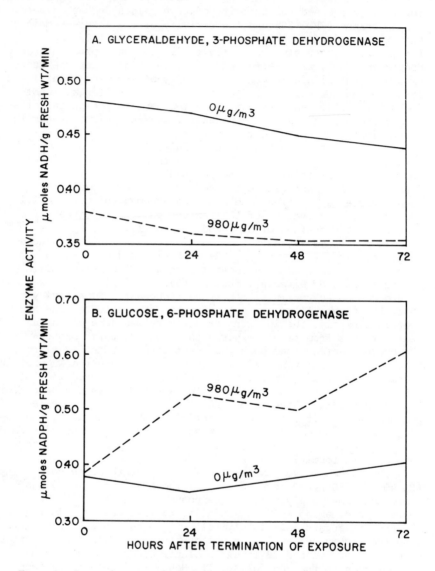

*Figure 3.   Ozone alteration of dehydrogenases active in sugar oxidation in soybean leaves. Each mean is based on eight observations.*

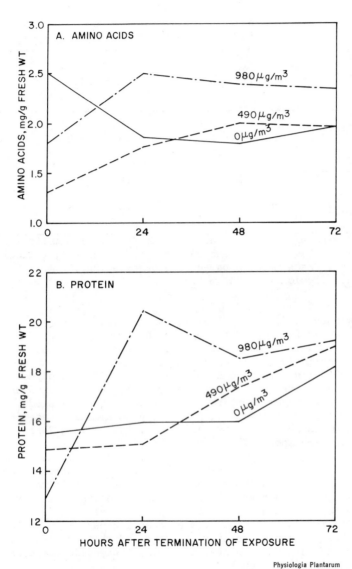

Physiologia Plantarum

*Figure 4. Effect of ozone on foliar levels of amino acids and protein in soybean leaves (3). Each mean is based on four observations.*

foliar protein.

Table III.  Effects of Ozone on Nitrate Reduction
in Soybean Leaves[1]

| NR  activity<br>($NO_2^-$ μmol/g fresh wt/hr) | Ozone conc.<br>(μg/m³) | |
|---|---|---|
| | 0 | 980 |
| In vivo NR<br>    Leaf Disc | 18.0 | 7.3 |
| In Vitro<br>    NADH dependent NR | 9.0 | 8.5 |
| In Vitro coupled assay<br>    NADH dependent NR | 3.3 | 1.4 |

Plants were exposed to ozone for 2 hr and the 1st trifoliate
leaves were assayed for enzyme activity immediately following
exposure.  Each mean is based on 9 observations.
Source:  Physiol. Plantarum (3).

Preliminary studies with field-grown alfalfa exposed to chronic
ozone levels (200 μg/m³ throughout the growing season) indicated
that the levels of amino acids and protein were increased 19 and
28% respectively (Neely and Tingey, unpublished).
     The activities of enzymes associated with phenol metabolism
in soybean leaves can be altered by a single ozone exposure
(0 or 980 μg/m³ of ozone for 2 hr) (Tingey, unpublished).  Immediately following exposure, phenylalanine ammonia lyase (PAL)
activity was suppressed by ozone (Fig. 5).  Within 24 hr, the
PAL level in exposed leaves was significantly above that of the
control.  The activity of polyphenol oxidase (PPO) is suppressed
immediately following ozone exposure (Fig. 5).  After 24 hr, PPO
activity in exposed leaves was greater than that in the control
and remained at a higher level for the duration of the experiment.  PAL is a key exzyme in phenol biosynthesis and its
activity is usually associated with an increase in phenol levels.
The increased enzyme activities suggested that ozone-exposed
leaves should contain elevated levels of phenolic compounds and
phenol oxidation products.
     Ponderosa pine seedlings exposed to a chronic ozone dose
(200 μg/m³, 6 hr/day throughout the growing season) also exhibited changes in phenol metabolism (Tingey and Wilhour, unpublished).  Between the first and second months of exposure,
total phenol concentration began to increase in foliage of exposed Ponderosa pine seedlings (Fig. 6).  Throughout the remain-

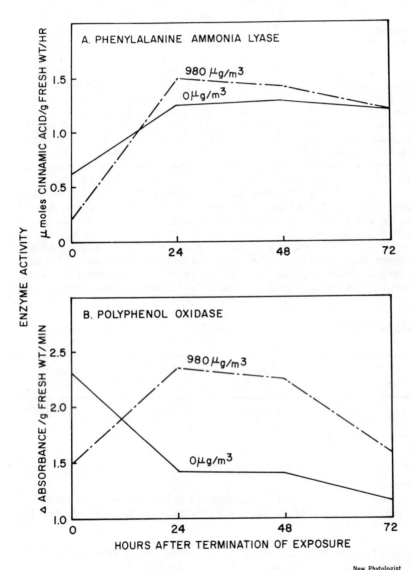

*Figure 5.  Ozone alteration of PAL and PPO activity in soybean leaves (8). Each mean is based on eight observations.*

der of the growing season, phenol levels in exposed plants exceeded levels in the controls. This increase in phenol concentrations was associated with an increase in the level of soluble
carbohydrates (Fig. 2) in exposed plants. The concurrent increase in soluble carbohydrate and phenols suggested that ozone
may have stimulated glycoside formation in Ponderosa pine
foliage.

Ozone treatment is known to increase glycoside and total
phenol levels in plants (16, 17, 18, 19) The increase in enzymes
of phenol metabolism and in glycosides suggests that some of
the pigments observed in ozone-injured tissue are phenolic
products. Howell and Kremer (18) have isolated a reddish brown
pigment complex from ozone treated bean leaves that contained
caffeic acid, amino acids, sugars and some metals. The ozone-
induced synthesis of glycosidic pigments would also reduce the
amount of photosynthate available for translocation and growth.

The effects of ozone are not limited to altering metabolic
pools in plant foliage; they can also induce the formation of
toxic compounds. Leachate was collected from fescue leaves 2
weeks following a 2 hr exposure to 590 $\mu g/m^3$ ozone and used to
irrigate clover plants (20). Top and root dry weights of ladino
clover were unaffected by leachate from ozone-treated fescue
leaves. However, the leachate from the ozone-exposed fescue
leaves reduced the nodule number of ladino clover (Table IV).

Table IV.  Effect of Fescue Leaf Leachates on the Growth
of Ladino Clover

| Treatment | Top dry wt (g) | Root dry wt (g) | Nodule number |
|---|---|---|---|
| Fescue leachate | 1.08 | 0.27 | 169 |
| Ozone treated fescue leachate | 1.02 | 0.24 | 78 |

The clover plants were 1 month old when the applications of
fescue leaf leachates were initiated. The leachates were applied
to the soil daily for 1 month when the cover was harvested.
Each mean is based on 10 observations. Data from Kochhar (20).

The leachate from ozone-exposed leaves could affect plant-plant
interactions and influence the population of soil microorganisms.
These metabolic alteration in plant foliage could also modify
litter turnover rates.

## Changes in Metabolite Pools in Roots

The observed ozone-induced growth reductions of roots could result from (a) a direct toxic effect of ozone on the root, (b) an ozone modification of the foliage metabolism which alters the quantity and/or quality of metabolites translocated to the roots, or (c) an alteration in soil chemistry.

Several factors suggested that the observed reductions in root growth were the result of ozone altering foliage metabolism rather than of its direct effect on the root. Ozone is very re-active and would be unlikely to penetrate into soil that is either moist or contains organic matter. An experiment was con-ducted to determine if air containing ozone could be drawn through soil columns (Blum and Tingey, unpublished). An air stream containing 980 µg/m$^3$ ozone was drawn through soil columns and then analyzed to determine the ozone concentration in air leaving the columns. During the 2 hr exposure period, 120 µg of ozone was applied to soil columns with these characteristics: a surface area of 78.5 cm$^2$ and 2 or 4 cm deep; containing gravel, sand, jiffy mix or jiffy mix: gravel (1:2, V/V). However, no ozone could be detected in the air stream exiting the columns. To determine the depth that ozone could diffuse into the soils, reduced 2, 6-dichlorophenol-indophenol was impregnated onto sand. The sand was placed in an atmosphere containing 980 µg/m$^3$ ozone for 2 hr and the depth of ozone penetration was measured as the depth that the reduced dye on the sand was oxidized. By this measure, ozone penetrated less than 20 mm into sand (Blum and Tingey, unpublished).

The ability of ozone to indirectly alter root growth can be shown by the use of root exudates. Root exudates from fescue plants exposed to a single ozone dose (590 µg/m$^3$ for 2 hr) in-hibited the top and root growth of ladino clover and also reduced nodule numbers (Table V) ([20]).

Table V. Effects of Root Exudates from Fescue Plants Exposed to Ozone on the Growth of Ladino Clover

| Treatment | Top dry wt (g) | Root dry wt (g) | Nodule number |
|---|---|---|---|
| Fescue root exudate | 1.40 | 0.40 | 149 |
| Ozone treated fescue root exudate | 1.19 | 0.31 | 101 |

The fescue plants were 1 month old when exposed to 0 or 590 µg/m$^3$ ozone for 2 hr. The 1-month-old clover plants were ex-posed to the fescue root exudates for 1 month and then harvested. Means are based on 10 observations. Data from Kochhar ([20]).

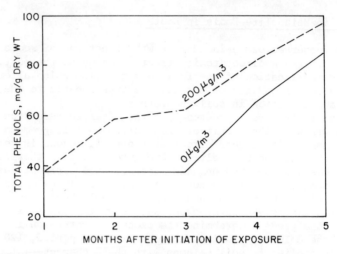

Figure 6.   *The influence of chronic ozone exposure on the levels of total phenols in the foliage of seedling Ponderosa pine. Each mean is based on nine observations.*

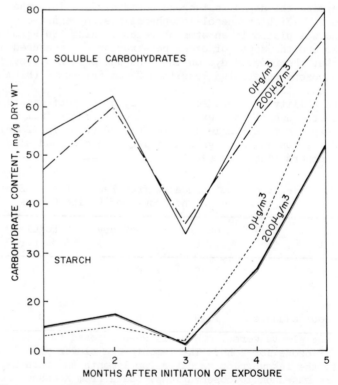

Figure 7.   *The effect of chronic ozone exposure on the levels of carbohydrates in the roots of Ponderosa pine seedlings. Each mean is based on nine observations.*

Ozone either induced the fescue plants to form a phytotoxin or caused some metabolite necessary for growth or nodulation to be removed from the root exudate stream. In either case, the ozone exposure of the foliage influenced the metabolite pools in the root. Root exudates from ozone-exposed plants could also influence the type or number of soil microorganisms in the rhizosphere and alter plant-plant interactions.

The levels of stored carbohydrates in Ponderosa pine roots were reduced by chronic ozone exposure (200 µg/m$^3$, 6 hr/day throughout the growing season) (Tingey and Wilhour, unpublished). The level of soluble carbohydrates in roots of Ponderosa pine seedlings exposed to ozone tended to be less than those in the control level throughout most of the season (Fig. 7A). Starch levels in the roots of exposed and control plants were similar until fall when starch storage began. At that time, the exposed plants accumulated starch at a significantly slower rate than the controls (Fig. 7B). This reduction in food reserves could hinder the initiation of growth the following spring.

Wardlaw (21) has indicated that the growth of established shoots appears to have priority over the growth of buds and roots when the assimilate is deficient. Roots and buds appear to be the poor relations among the plant organs, receiving only the photosynthate in excess of the requirements of other parts. The ozone-induced depression of photosynthesis coupled with a retention of carbohydrates in the foliage and a reduction in translocation of photosynthate to the roots could explain the reduction in both storage metabolites and root growth.

Levels of metabolites other than carbohydrates were also altered in roots of Ponderosa pine seedlings as a result of chronic ozone exposure (Tingey and Wilhour, unpublished). After three months of exposure, which coincided with the appearance of visual injury, amino acid levels in roots of exposed seedlings were elevated above those of the controls (Fig. 8A). The amino acid levels remained above that of the control for the remainder of the season. Also the levels of Kjeldahl nitrogen were higher in roots of exposed than of control plants throughout the season (Fig. 8B).

Alterations in root metabolism as a result of ozone treatment are also reflected in the nodulation of soybeans (21). When soybeans were exposed to a single acute ozone dose (1470 µg/m$^3$ ozone for 1 hr) and harvested at weekly intervals following exposure, nodule number increased more slowly in exposed plants (2 nodules/wk) than in control plants (8 nodules/wk) (Fig. 9). A similar trend was seen in the nodule weight per plant and leghemoglobin content per nodule was not affected. This indicated that the effect of the ozone treatment was to reduce nodule number; the reduction of nodule number caused a concurrent reduction in total nodule weight and leghemoglobin per plant. The decrease in leghemoglobin, an indicator of nitrogen fixation capacity (22) suggested the nitrogen fixation would also be

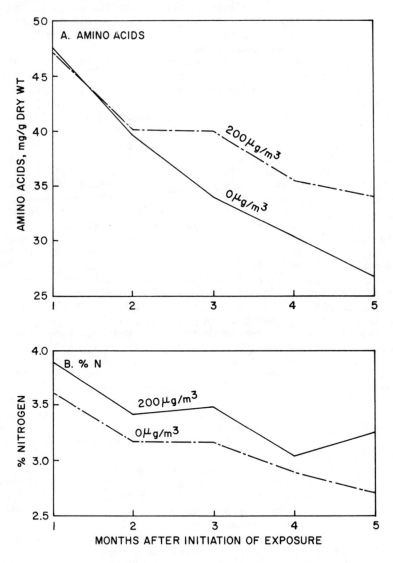

*Figure 8. Ozone alteration of amino acid and Kjeldahl nitrogen levels in the roots of Ponderosa pine seedlings exposed to chronic ozone doses. Each mean is based on nine observations.*

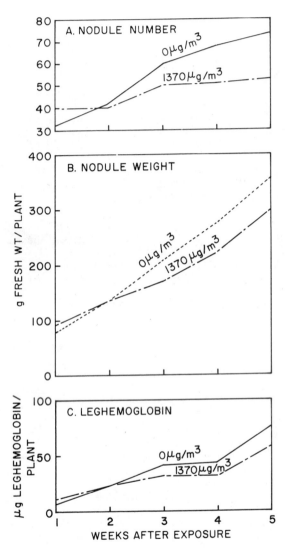

*Figure 9.   Effect of a single ozone dose on the nodulation parameters of soybean. Each mean is based on 10 observations.*

reduced by the ozone treatment. Soybeans exposed to 490 µg/m$^3$ ozone for 3 hr/day, 3 days/week also exhibited a reduction in nodule number and weight similar to that resulting from an acute exposure (Weber, unpublished).

## Literature Cited

1. Black, M. and Edelman, J. "Plant Growth." Heineman Educational Books, Ltd. London. (1970).

2. Tingey, D. T., Dunning, T. A. and Jividen, G. M. A154-A156. "Proceedings Third International Clean Air Congress." VDI-Verlag, GmbH, Dusseldorf, Federal Republic of Germany. (1973).

3. Tingey, D. T., Fites, R. C. and Wickliff, C. Physiol. Plantarum (1973) 29:33-38.

4. Hill, A. C. and Littlefield, N. Environ. Sci. Tech. (1969) 3:52-56.

5. Miller, D. R., and Parmeter, J. R. Jr., Flick, B. H. and Martinez, C. W. J. Air Pollution Control Assoc. (1969) 19:435-38.

6. Bennett, J. H. "Effects of ozone on leaf metabolism." Ph.D. Thesis University of Utah, Salt Lake City, Utah. (1969).

7. Barnes, R. L. Can. J. Bot. (1972) 50:215-19.

8. Tingey, D. T., Fites, R. C. and Baharsjah, J. New Phytologist (1974) 73:21-29.

9. King, M. E. "Biochemical Effects of Ozone." Ph.D. Thesis Illinois Institute of Technology, Chicago, Illinois. (1961).

10. Chow, C. K. and Tappel, A. L. Archives of Environmental Health (1973) 26:205-208.

11. Goodman, R. N., Kiraly, Z. and Zaitlin, M. "The Biochemistry and Physiology of Infectious Plant Diseases." D. Van Norhand Company, Inc., Princeton, New Jersey. (1967).

12. Wood, R. K. S. "Physiological Plant Pathology." Blackwell Scientific Publications, Oxford, England. (1967).

13. Klepper, L., Flesher, D. and Hageman, R. H. Plant Physiol. (1971) 48:580-590.

14. Ting, I. P. and Mukerji, S. K. Amer. J. Bot. (1971) 38:497-504.

15. Tomlinson, H. and Rich, S. Phytopathology (1967) 57:972-974.

16. Koukol, J., and Dugger, W. M. Jr. Plant Physio. (1967) 42:1023-1024.

17. Menser, H. A., and Chaplin, J. R. Tobacco Science (1969) 13:169-170.

18. Howell, R. K. and Kremer, D. F. J. Environ. Quality (1973) 2:434-38.

19. Tomlinson, H. and Rich, S. Phytopathology (1973) 63:903-906.

20.  Kochhar, M.  "Phytotoxic and competitive effects of tall
     fescue on ladino clover as modified by ozone and/or
     Phizoctonio solani."  Ph.D. Thesis.  North Carolina State
     University. Raleigh, NC. (1973).
21.  Wardlaw, I. E. Bot. Rev. (1968) 34:79-105.
22.  Graham, P. H. and Parker, C. A.  Austral. J. Sci. (1961)
     23:231-32.
23.  Tingey, D. T., Heck, W. W. and Reinert, R. A.  J. Amer.
     Soc. Hort. Sci. (1971) 96:369-371.

# 5

# The Role of Potassium and Lipids in Ozone Injury to Plant Membranes

ROBERT L. HEATH, PHROSENE CHIMIKLIS, and PAULA FREDERICK

Departments of Biology and of Biochemistry, University of California,
Riverside, Calif. 92502

Low ozone concentrations damage green plants by mechanisms which remain obscure. The effects of ozone on cell constituents are multiple and diverse, ranging from changes in ribosomal formation (1) and soluble protein levels (2, 3), to altered sulfhydryl content (4). In many studies the changes observed are not immediate, but take hours and sometimes even days to develop. However, the changes which are immediately noted can usually be related to water loss from the leaf. In fact, a commonly noted gross visible symptom is a necrosis or a dessication-induced collapse of the palisade layers (5). In many cases, acute injury results in closure of the stomata which prevents ozone penetration and further water loss (6).

Several major features of ozone injury in pinto bean (Phaseolus vulgaris) have been noted by Dugger, Ting and their co-workers (5) and many of these findings apply in other plant systems as well. In sum, their research has shown that the extent of ozone sensitivity is dependent upon (a) leaf age, (b) photoperiod, (c) plant water potential and (d) relative content of soluble sugars and amino acids (organic acids have not been examined). Furthermore, permeability changes are commonly noted in ozone-treated tissues. Evans and Ting (7) have shown that ozone exposure of plants results in a decline in $Rb^{86}$ uptake and that a loss of $Rb^{86}$ occurs in preloaded leaf discs cut from the plants soon after exposure. Perchorowicz and Ting (8) have also shown changes in glucose permeability in ozone-exposed bean leaves.

It is our hypothesis that the diverse responses of green plants to ozone injury which have been measured previously are due to ionic imbalances induced by ozone. More specifically, ozone appears to induce gross leakage of $K^+$ out of plant cells by some reaction in the membrane, and abolishes uptake of this ion by inhibition of a "$K^+$ pump" located on the plasmalemma. Ozone interactions with the membrane are believed to involve both critical sulfhydral groups and fatty acid residues and they may cause changes in the membrane's fluidity.

Ozone is a very reactive gas and is highly soluble in water (9); consequently, it seems unlikely that ozone could avoid reacting with the plasmalemma as it passes into the cell. Coulson and Heath (10) have demonstrated such reactivity using isolated chloroplasts (Table I). Ozone bubbled into an aqueous solution containing isolated grana stacks inhibits electron transport, as measured by the reduction of ferricyanide. Similar doses of ozone also inhibit bicarbonate-stimulated oxygen evolution in intact chloroplasts. If, after ozone treatment, the intact plastids are osmotically ruptured and the grana are released, the electron flow in these grana is unaffected (Table I,C). These observations indicate that ozone must either alter the outer membrane of the intact plastid or interact with the $CO_2$-fixing enzymes in the stroma. An alteration in permeability of the chloroplast's membranes has been demonstrated by Nobel and Wang (11), who noted that the reflection coefficient of several organic molecules was markedly lowered in plastids exposed to ozone.

The inability of the plasmalemma of ozone-treated cells to retain $K^+$ was first noted by Chimiklis and Heath (12). A cation-specific electrode coupled to an antilog converter was used to change the "Nerst-equation" response of the electrode output into a linear recording. A typical response is shown in Figure 1. Experiments are performed at $38°C$ in 10mM Tris-Cl, 1 mM $CaCl_2$, at pH 8 or 9 using Chlorella sorokiniana (strain 07-11-05, a thermophilic strain). After 10 to 20 min of suspension, the rate of $K^+$ loss becomes constant. The apparent "slow leak" of $K^+$ out of the cell was shown to indicate an exchange of $K^+$ inside the cell with Tris outside (13), possibly similar to the exchange mechanism described by Barber and coworkers (14). As shown here, oxygen bubbled into the system has no apparent effect upon this leak. The loss continues for more than three hr, finally ceasing upon reaching a steady-state level of external and internal $K^+$. Ozone bubbled into this system causes a rapid increase in the rate of $K^+$ loss; the steady-state rate is variable and dependent on the influx and efflux of $K^+$ (described later). It is important to note that the ozone-induced leak occurs immediately (within tens of seconds).

Figure 2 depicts the temperature dependence of the "control" $K^+$ leakage and ozone-induced $K^+$ leakage as an Arrhenius plot. The slope of both lines is about 15 kcal/mole, indicating the same membrane-mediated efflux mechanism. Ozone does not change the energy of activation, but merely stimulates the efflux by a factor of nearly 8 at all temperatures. This suggests that ozone does not induce a random disruption or a disintegration of the membrane.

At $38°C$ the high leakage rate ceases rapidly, but not immediately, upon cessation of ozone bubbling (taking 5-7 min to return to the control level). At pH 8 or 9, the ozone

Table I. Inhibition of electron flow in isolated chloroplasts
by ozone

---

A.  Isolated Grana Stacks (Ferricyanide Reduction)

| Preparation Age (min) | $O_3$ dose (nmoles) | % Inhibition |
|---|---|---|
| 20-60 | 80 | 20 ± .4 |
| " | 220 | 30 ± 2 |
| " | 490 | 40 ± 2 |

B.  Intact Plastids (Bicarbonate-Stimulated $O_2$ Evolution)

| Preparation Age (min) | $O_3$ dose (nmoles) | % Inhibition |
|---|---|---|
| 20-50 | 400 | 36 ± 8 |
| 60-100 | " | 47 |
| 110-145 | " | 31 ± 12 |

C.  Osmotically Ruptured Intact Plastids (Ferricyanide Reduction)

| Preparation Age (min) | $O_3$ dose (nmoles) | % Inhibition |
|---|---|---|
| 20-60 | 80 | 3 ± 3 |
| " | " | 9 ± 3 |
| " | " | 3 ± 2 |

---

(± represents either S.D. for A and C or variation for B.)


These data are from Coulson and Heath (10). The $O_3$ doses were
delivered for short times (5 min in Part A, and 1 min in Parts
B & C) to prevent breakage of the plastids. In Part C, the
intact plastids were treated as in B and then osmotically
checked by resuspension in a solution containing no osmoticium,
before assaying as in Part A. Ferricyanide reduction was
measured spectrophotometrically.

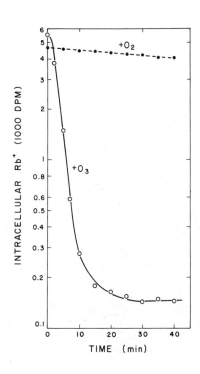

*Figure 4. The transport of ⁸⁶Rb by Chlorella under O₃ stress A. Control efflux from pre-loaded cells. A culture of 38°C-grown C. sorokiniana was preloaded with ⁸⁶Rb for 24 hrs. Cells were concentrated by centrifugation, washed, and resuspended as described in Figure 1, and assayed for ⁸⁶Rb loss by millipore filtration. The cells were dried, bleached, and counted by liquid scintillation as described by Fredrick and Heath (24). B. Influx into cells. A culture of exponentially growing C. sorokiniana was centrifuged, washed, resuspended in the Tris-Cl solution (see Figure 1) plus 100 μM KCl, and placed in a 38°C water bath. At time intervals after ⁸⁶Rb addition, samples of cells were assayed as described in A.*

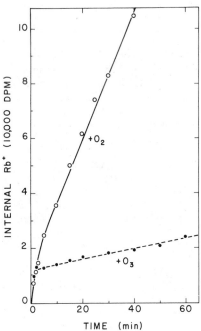

remaining in solution is rapidly broken down (9) and, thus, is
not involved in the recovery of lower leakage rates. At lower
temperatures, not only is the rate of the ozone-induced leakage
lower, but the return to control level is also slower. This is
shown in Figure 3A where the ozone-induced leakage rate is
normalized (to unity) for all temperatures. The control rate
is about 0.12 at all temperatures and the increasing time required for return to the control rate at lower temperatures can
be easily seen. Figure 3B shows the temperature dependence of
the half-time of the $K^+$ leak cessation process, again as an
Arrhenius plot. The slope of the line indicates an energy of
activation of about 16 ± 2 kcal/mole, with a possible dis-
continuity at 15°C (also observed in Fig. 3A where the rate of
ozone-stimulated $K^+$ leak does not return to the control level).
This energy of activation is high enough to indicate the
operation of an active repair mechanism and not just a physical
process.

These $K^+$ measurements have monitored net $K^+$ flux across
the membrane. The above data can be more easily interpreted if
$Rb^{86}$ is used as a unidirectional tracer of $K^+$ movement in
Chlorella. Figure 4A shows a "leak-out" or efflux experiment
with Chlorella previously loaded with $Rb^{86}$ overnight (for 6+
generations). These cells are washed and suspended in a
solution of Tris-Cl at 38°C, as above. The control, bubbled
with oxygen, shows little loss of $K^+$ (nearly linear with a
20 dpm loss/min). With the addition of ozone to the solution,
a rapid loss of intra-cellular $Rb^{86}$ occurs (exponential loss,
initial rate= 1000 dpm loss/min), or the unidirectional efflux
is increased by ozone by a factor of nearly 50. The external
$Rb^{86}$ concentration reaches steady-state at about 20-25 min
when the concentration of external $K^+$ is nearly equal to that
of internal $K^+$, as measured by the $Rb^{86}$ specific activity.

When ozone is turned-off at some point during this rapid
loss period, the high rate of $Rb^{86}$ leakage ceases within
2-3 min and returns to the control leakage rate. Thus, this
permeability change, measured by efflux, appears to be
reversible, as is net $K^+$ leakage measured with the cation electrode.

The influx of $K^+$ is thought to be controlled by a "pump"
(15). Figure 4B shows the influx of $Rb^{86}$ (as a $K^+$ tracer)
with normally-grown cells. The $Rb^{86}$ is placed in the standard
external medium with 100 uM KCl. After a small, rapid uptake
of $Rb^{86}$ into the cell--thought to be due to $Rb^{86}$ binding to
the cell wall or lodging within the apparent free space
(15, 16) --the rate of increase in intracellular $Rb^{86}$ (or $K^+$)
is linear ($10^5$ DPM/30 min) for cells bubbled with oxygen.
However, ozone bubbling inhibits the $Rb^{86}$ uptake nearly completely (5% of the control rate = 5 x $10^3$ DPM/30 min). The
effect of ozone on the influx, unlike the effect on leakage, is
irreversible over at least 30 min. Turning-off the ozone after

5 min of bubbling does not result in restoration of the influx rate to the control level. Thus, while the influx is irreversibly inhibited (nearly completely) by ozone, the efflux is reversibly increased (greatly) by it. The influx of $Rb^{86}$ is also inhibited by sulfhydryl reagents such as $Hg^{++}$ and p-chloromercuribenzoate (data not shown). The efflux of $Rb^{86}$ has not been investigated completely, but the $K^+$ loss at $38°C$ measured by the cation-specific electrode is not altered much either by antibiotics such as valinomycin or nigericin, or by metabolic inhibitors such as $N_2$, $CO$ or DNP.

These effects of ozone on the permeability of this algal system may be summed up as follows: during exposure, $K^+$ leaks out of the cell and cannot be pumped back in; even when exposure is discontinued and the $K^+$ leak has ceased, the cell is still unable to restore the lost $K^+$; if a critical level of cellular $K^+$ has been lost at this point, turgor pressure and water content become depressed, the metabolic control provided by ionic balance is lost and general metabolic alterations should be observed ([17]).

Studies on the growth of Chlorella following ozone treatment lend general support to the sequence of events outlined above. Cells treated with ozone for 10 min under the above conditions lose $10^{-15}$ moles of $K^+$/cell (out of $5.3 \times 10^{-15}$ moles/cell), do not resume growth for at least 10 hr and become bleached when placed in normal growth medium. Untreated cells resume optimal growth patterns within 2-3 hr.

By way of comparison with higher plant systems, it should be noted that the amount of $K^+$ in the leaf (measured in Phaseolus vulgaris) reaches a maximum (on a fresh weight basis) at approximately the same time as the leaf's maximal sensitivity to ozone ([2,5]; Fig. 5). This period occurs when the leaf expansion rate is greatest. During the developmental period when high turgor pressures are thought to be required for expansion of walled cells ([18]), high intracellular $K^+$ levels would be expected. Dugger et al. ([19]) have shown that leaf soluble sugar levels are minimal at the period of greatest ozone sensitivity. It appears, therefore, that turgor pressure within the left cell may alternately derive from either high $K^+$ or high soluble sugar (+ organic acids?) levels. Furthermore, our preliminary data supports previous data ([20]) which shows a diurnal fluctuation of leaf $K^+$ (with a maximum at about 4-5 hr following the start of the day period, which also corresponds to the diurnal period of maximum ozone sensitivity ([2])).

It is our hypothesis that the presence of large amounts of $K^+$ in the cell is a major determining factor in ozone injury at the cellular level. The alteration of $K^+$ permeability (coupled with an inhibition of the mechanism for regaining the lost $K^+$) causes a large loss of $K^+$ (down an electro-chemical potential gradient) followed by a rapid loss of osmotic water. It is ultimately this loss of osmotic water which leads to dehydration and ionic imbalances within the cell. We feel it is these events which trigger the subsequent metabolic changes which have been observed over the last decade (5).

A full understanding of the injury process should also include knowledge of the initial points of ozone attack. Two biochemicals -- unsaturated fatty acids and sulfhydryls -- are generally regarded as primary sites of ozone injury (4, 21). To better study the possible reactions of these compounds, a differential ozone-uptake instrument was constructed (Fig. 6). In this system, the ultraviolet absorbance of ozone provides a method for quantifying ozone uptake by aqueous solutions. Pure dry $O_2$ gas passes through a flow meter into an ozone generator consisting of ultraviolet lamps in an aluminum cylinder (10). The mixture of ozone and oxygen passes through a valve A which directs the flow through an activated charcoal filter (Scrub I) to remove ozone for control samples. The flow then passes through a 10 cm path length quartz cell in the sample compartment of a Cary Spectrophotometer (Model 15) and then through valve B which directs it into the reaction vessel. Unreacted gas passes out of this vessel and through valve C which can direct the gas through a second activated charcoal filter (Scrub 2) to remove ozone. The flow finally passes through a second quartz cell in the reference compartment of the spectrophotometer and out through a vent.

The typical trace obtained with this instrument (Fig. 7) can be used to explain the operation of the system. The absorbance recorded by the spectrophotometer with both Scrubs turned in is shown as a function of time (the right-hand ordinate is converted to ppm ozone by using an extinction coefficient (22)). Scrub 1 provides a base line for the absorbance (no ozone). At 2 min, Scrub 1 is removed and the absorption rises to a level calculated to be 250-260 ppm of ozone. A 30 second lag occurs due to the volume of the system. The reaction vessel is added to the flow route at 4 min, and at 8 min Scrub 2 is removed. The absorbance value now recorded is the amount of ozone absorbed by the system between the sample and reference cuvettes. Most of this is due to breakdown of ozone in the alkaline solution contained in the reaction vessel. If this solution is replaced by 1 N NaOH, absorbance at this point is much higher; when the solution is 0.1 N HCl, absorbance is very low. This corresponds with the known rate of ozone breakdown as a function of pH (9). If linolenic acid (Sigma chemical, 99% pure, C18:3 (9c, 12c, 15c)) in acetone is added to the reaction medium (+ in the figure), the absorbance rises rapidly to a relatively high value of 210 ppm, indicating that most of the ozone is removed by the linolenic acid solution. After 2 to 3 min, this high absorbance declines slowly to a lower value. A low steady-state absorbance is reached in about 12-15 min. In order to convert this recording to a rate of ozone breakdown, the difference in absorbance between with (+) and without (-) added linolenic acid is

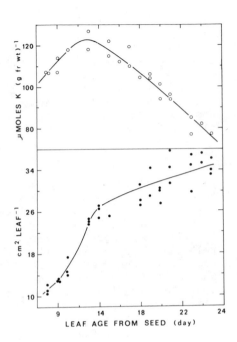

Figure 5. Age-dependent concentrations of K⁺ within the primary leaves of Phaseolus vulgaris L. var. red pinto. The concentration of $K^+$ ($\bigcirc$—$\bigcirc$) observed in primary leaves of Phaseolus vulgaris at successive ages from seed. Leaves were measured for area ($\bullet$—$\bullet$), washed with 10 mM $CaCl_2$, digested with $HNO_3$, and $K^+$ was measured (after neutralization of the solution) with a cation electrode. Growth conditions were as previously described (7, 8).

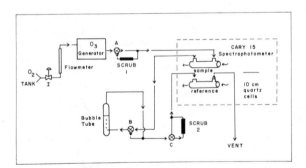

Figure 6. Schematic of apparatus measuring $O_3$ uptake. Ozone was generated by the flow of the carrier gasses over UV lamps, as in (10). The scrubs (1 and 2) are bypass tubes filled with activated charcoal used to decompose $O_3$. Valves A, B, and C provide different paths for the gas flow. The bubble tube contained 50 ml of solution and has been previously described (24).

multiplied by the flow rate of air to obtain ml $O_3$/min decomposed.

The concentration of ozone taken up by the media containing linolenic acid is plotted against time after addition in Figure 8. The rate of ozone breakdown is constant (ozone uptake linear with time) for the first two min until about 0.12 ml ozone are absorbed and then the rate decreases sharply, reaching a steady-state rate of ozone uptake between 10-12 min. This first break in the curve corresponds to an ozone uptake of 0.12 ml + (24 moles/liter) = 0.005 millimoles (or $10^{-4}$ M). This is equivalent to 1 mole of linolenic acid added per mole ozone absorbed. Thiobarbituric acid reactant production is also plotted on the same axis. This compound (TBA reactant) probably arises by formation of a three-carbon fragment (malondialdehyde) from the ozone-induced oxidation of linolenic acid (23). The rate of TBA reactant formation is also linear for the first 2 min at which point the curve undergoes a less pronounced break. Malondialdehyde formation ceases immediately when the ozone is shut off (Scrub 1 on). An oxygen control sample produced no malondialdehyde.

The amount of TBA reactant can be converted into moles of malondialdehyde by the extinction coefficient of 155 $mM^{-1}cm^{-1}$ ozone (22), and the yield of this reaction, or ratio of malondialdehyde/ozone taken-up, is shown in Figure 9. Notice that the yield for ozone-treated linolenic acid varies with time of reaction from about 3% to over 30%. These results differ from those for lipid peroxidation reactions which also give rise to malondialdehyde but have yields of 2-5% (23).

Malondialdehyde formation can be used to help characterize the primary site of ozone attack in the Chlorella system. Frederick and Heath (24) have demonstrated that Chlorella subjected to ozone lose their viability exponentially with exposure time. Cell death is determined by the ability of the cells to form colonies on agar plates supplemented with glucose. In Figure 10, the viable cells are shown as a function of time of ozone exposure. After about 5 min cells begin to die, and after 20 min of exposure less than 2% of the original cells are viable. Ozone uptake (measured as described above) is also shown in this figure. Much less than 0.1 μmoles of ozone (<5 x $10^{-17}$ moles/ cell) can be observed to be absorbed when the cells are alive, but with the loss of viability, ozone uptake begins. The same figure shows the production of malondialdehyde, which also begins as the cells begin to lose viability. The kinetics for both ozone uptake and malondialdehyde production seem to be exponential and are nearly the inverse of the viability decline.

The kinetics of these processes are tabulated in Table II, for exposure times of 10, 15 and 20 min. The ratio of malondialdehyde to ozone uptake is high at first (9%) and then declines. The kinetics of these ratios are different than those for ozone reactions with free fatty acids, but the ratio is of

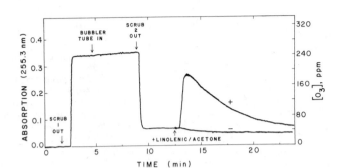

*Figure 7. Trace of an $O_3$ uptake measurement. The uptake of $O_3$ by the instrument (see Figure 6) and by a solution in the bubble tube is measured by the UV spectral absorption difference. The ppm of $(O_3)$ concentration was calculated based on an extinction coefficient of 138 atm$^{-1}$ (22). Scrub out refers to a valve change resulting in the removal of the indicated scrub from the system. Linolenic acid (100 $\mu M$, final concentration in 0.5 ml acetone) is added where indicated 0.5 ml (+). The total solution in the bubble tube is 50 mM Tris-Cl pH 8.2 (50 ml). An injection of acetone alone provided the control trace (−).*

*Figure 8. The kinetics of $O_3$ uptake and Thiobarbituric acid reactant produced by ozonolysis of linolenic acid. From the $O_3$ uptake described in Figure 7, the amount of $O_3$ is calculated by integrating the area under the +1-linolenic curve (in ppm/min) and multiplying by the air flow rate (ml/min). The Thiobarbituric acid (TBA) reactant assayed according to Heath & Packer (23), is given as absorbance difference (A 532–A 580).*

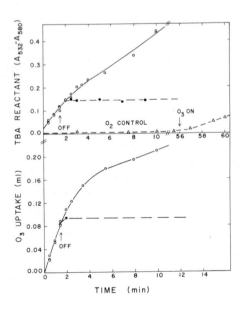

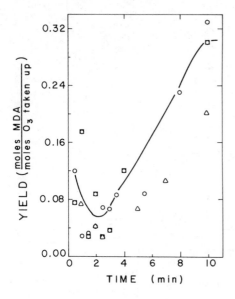

*Figure 9. Stoichiometry between malondialdehyde formation and $O_3$ uptake. The data from Figure 8 was used to construct this curve, by converting ml of $O_3$ taken up into moles by the perfect gas volume at 25°C (24 l/mole) and by converting the TBA reactant value into malondialdehyde formed by using an extinction coefficient of 155 $mM^{-1}$ $cm^{-1}$ (23). Different symbols refer to separate experiments.*

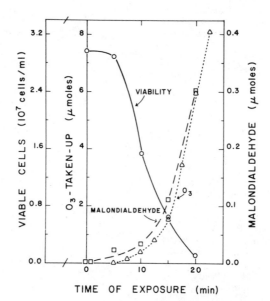

*Figure 10. Production of malondialdehyde, the change in* Chlorella *viability and uptake of $O_3$ by a suspension of* Chlorella *cells. A sample from a culture of 38°C grown C. sorokiniana var. pacificensis ($3 \times 10^7$ cells/ml autotrophic medium) was treated with 180 ppm ozone. Thiobarbituric acid reactants were assayed by the method of Heath & Packer (23), viable cells by plating on glucose-supplemented agar medium, and ozone uptake on a Cary spectrophotometer as described in Figures 6–8.*

the same order of magnitude.  The ozone absorbed per non-viable
cell is low at first and then rises.  At the lowest it corresponds to about $3 \times 10^8$ molecules ozone broken down per killed-
cell, nearly that observed for E. coli by McNair-Scott and
Lesher (25).


Table II.  Kinetic analysis

| $O_3$ Exposure time (min) | Malondialdehyde ( mole ) / ozone uptake ( mole ) | Ozone uptake per non-viable cell (moles $O_3$/dead cell) |
|---|---|---|
| 10 | 9.1% | $5 \times 10^{-16}$ |
| 15 | 7.1% | 13 |
| 20 | 5.1% | 42 |

These ratios are from calculations of the data in Fig. 10.


The ozone-induced formation of malondialdehyde should also
be observable as a loss of unsaturated fatty acids from Chlorella
(24) (Table III).  Table III shows the average percentage of
total fatty acid concentrations for the five major fatty acids
found in this alga.  Column four shows the changes in the fatty
acid after cells are exposed to ozone for 50 min (expressed as
a ratio of fatty acid in ozone-exposed cells to fatty acid in
the oxygenated-control cells).  The last column gives the confidence level as determined by variance analysis.  It is highly
probable that the amount of C16:0 increases (in percentage
concentration) with ozone exposure, while C16:3 decreases.  If
we assume that C16:0 actually remains constant, but that the
other fatty acids fall in concentration, the decrease in C16:3
and 18:3 levels is seen to be considerable.  It can be calculated
then that the total apparent loss of unsaturated fatty acids
(even the non-significant loss) is only about 3% of the total
fatty acid within the cell (0.075 out of $2.5 \times 10^{-15}$ moles fatty
acid/cell).  From the total loss of triunsaturated fatty acids
it can be calculated that about 20% of the triunsaturates are
converted into malondialdehyde.  Though these data do not define
the primary site of ozone injury as fatty acids, they do serve to
indicate the lethal possibilities of oxidation of a relatively
small percentage of these molecules.
     The second class of biochemicals which is significantly oxi-
dized by ozone is sulfhydryls.  The di-sulfhydryl compound,
dithiothreitol (DTT), readily loses 2 H to form a cyclic disulfide
(26).  This compound is used to study how ozone reacts with a
sulfhydryl system and whether these reactions are stoichiometric
(Fig. 11).

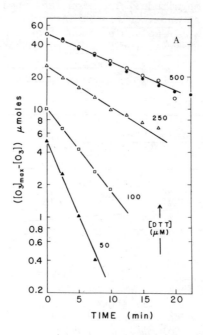

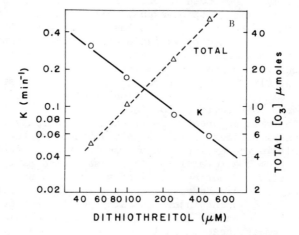

Figure 11.   The uptake of $O_3$ by a sulfhydryl reagent.
A. Semilogarithmic kinetic plot of $O_3$ uptake by Dithiothreitol (DTT).  The ozone uptake was calculated as
described in Figure 8.  Numbers at the side of the graph
refer to final concentrations of added DTT.  Different
symbols represent different experiments.   B. Log–log
plot of rate constant and total $O_3$ taken up with DTT
concentrations.  The rate constant (K) of $O_3$ uptake and
total ($O_3$) were calculated from Part A.

Table III.  Fatty acid changes induced by ozone

| Fatty acid | Average % concentration $+O_2$ | $+O_3$ | $O_3/O_2$ | Confidence level |
|------------|----------|--------|-----------|------------------|
| C16:0 | 25.4 | 28.6 | 1.14 ± 0.05 | >97.5% |
| C16:3 | 11.2 | 10.3 | 0.92 ± 0.04 | ~93% |
| C18:1 | 32.8 | 32.4 | 1.02 ± 0.04 | <75% |
| C18:2 | 5.4 | 4.8 | 0.90 ± 0.09 | <75% |
| C18:3 | 25.7 | 24.1 | 0.93 ± 0.05 | ~85% |

Chlorella sorokiniana var. pacificensis were treated with
180 ppm $O_3$ for 50 min in autotrophic media.  Lipids were extracted by using Chloroform/methanol and prepared for gas-liquid
chromatography (GLC) as described by Frederick and Heath (24).
The average % concentration of fatty acids were calculated from
3 GLC runs in 5 separate samples.  The $O_3/O_2$ column refers to
ratios of average % concentration and ± represents standard
deviation.  Confidence Level was calculated by least squares
analysis.

The data presented in Figure 11A depict ozone uptake with
increasing DTT concentrations in a semilogarithmic plot, since
it was observed that the $O_3$ uptake curve appeared to be exponential with time.  This plot shows (a) that ozone uptake is
logarithmic and (b) that the intercept (at zero time) and the
rate constant for the reaction (slope of lines) vary with the
amount of DTT added.  Figure 11B shows the variation of the
intercept (maximum $O_3$ uptake) and rate constant (K) as a function
of DTT concentration (as a log-log plot).  In this case, the
maximum $O_3$ uptake is proportional to twice the amount of added
DTT (slope of line on log-log plot is linear), and the rate
constant (K) varies with the inverse square root of DTT concentration (slope of line is - 1/2).  The conversion of DTT to the
cyclic disulfide should not give these kinetics, which (Fig. 11)
can only be generated if the following reaction occurs:

$$HS-R-SH + 2O_3 \longrightarrow HO-\overset{\overset{O}{\uparrow}}{\underset{\underset{O}{\parallel}}{S}}-R-\overset{\overset{O}{\uparrow}}{\underset{\underset{O}{\parallel}}{S}}-OH$$

Thus, it seems that since both sulfhydryls must be converted to
the sulfonate forms by ozone, Mudd's previous data (27) are
confirmed.
The speed of reaction of $O_3$ with linolenic acid and DTT
measured by this chemical system indicates that unsaturated fatty
acid and sulfhydryls react nearly equally if present in the
same concentration (100 µM) in an aqueous medium.  The questions
then are:  (a) are sulfhydryls and unsaturated fatty acids equally

accessible to ozone; and (b) what concentration of each are
present in cells?

   In conclusion, it is our view that $K^+$ plays a dominant role
in maintaining a favorable plant cell water status and that the
deleterious effects of ozone occur primarily due to disruption of
this normal ionic balance. We currently visualize the following
scheme of ozone damage: (a) ozone attack of an SH group or fatty
acid residue; (b) permeability changes of the membrane leading to
loss of both $K^+$ and osmotic water; (c) inhibition or loss of $K^+$
"pump" activity for reacquisition of lost $K^+$; (d) loss of a
critical level of ion; and (e) gross disruption of cellular metabolism.

   Acknowledgement. This project has been financed in part
with Federal funds from the Environmental Protection Agency under
grant number 801311. The contents do not necessarily reflect the
views and policies of EPA, nor does mention of trade names or
commercial products constitute endorsement or recommendation for
use. The authors would like to thank Ms. Joan Vereen for
technical assistance and Dr. Robert J. Beaver for statistical
analysis.

## Literature Cited

1.  Chang, C., Phytochem. (1971) 10: 2863-2868.
2.  Ting, I.P. and Mukerji, S.K. Amer. J. Bot. (1971) 58: 497-504.
3.  Tingey, D.T., Fites, R.C. and Wickliff, C., Physiol. Plant. (1973) 29: 33-38.
4.  Tomlinson, H. and Rich, S. Phytopath. (1968) 58: 808-810.
5.  Dugger, W.M. and Ting, I.P. Recent Advan. in Phytochem. (1970) 3: 31-58.
6.  Mansfield, T.A., Commentaries in Plant Sciences in Current Adv. in Plant Science 2: 11-20, Sciences, Eng. Med. and Business Data Ltd., Oxford. (1973).
7.  Evans, L.S. and Ting, I.P. Amer. J. Bot. (1973) 60: 155-162.
8.  Perchorowicz, J. and Ting, I.P. Amer. J. Bot. (1974) 61: 787-793.
9.  Kilpatrick, M.L., Herric, C.C. and Kilpatrick, M., J. Amer. Chem. Soc. (1956) 78: 1784-1789.
10. Coulson, C. and Heath, R.L. Plant Physiol. (1974) 53: 32-38.
11. Nobel, P.S. and Wang, C. Arch. Biochem. Biophys. (1973) 157: 388-394.
12. Chimiklis, P. and Heath. R.L. Plant Physiol. (1972) 495: 3.
13. Chimiklis, P. and Heath, R.L. Biochem, Biophys. Adv. (1974) (submitted).
14. Barber, J. and Shien, Y.J. J. Exp. Bot. (1972) 23: 627-636.
15. MacRobbie, E.A.C. Quart. Rev. Biophys. (1970) 3: 251-294.
16. Tromballa, H.W. and Broda, E. Arch. Mikrobid. (1972) 86: 281-290.
17. Stiles, W. and Cocking, E.C. "An Introduction to the Principles of Plant Physiology." pp. 297-299. Methuen and Co., London. (1969).
18. Cleland, R. Ann. Rev. Plant Physiol. (1971) 22: 197-222.
19. Dugger, W.M., Taylor, O.C., Cardiff, E. and Thompson, C.R. Proc. Amer. Soc. Hort. Sci. (1962) 81: 304-315.
20. Miller, E.C. "Plant Physiology," 2nd Ed., McGraw-Hill, N.Y. (1938) pp. 232-288.
21. Teige, B., McManus, T.T. and Mudd, J.B. Chem. and Phys. of Lipids (1974) (in press).
22. Inn, E.E.Y. and Tanaka, Y. Adv. in Chem. (1959) 21: 263-268.
23. Heath, R.L. and Packer, L. Arch. Biochem. Biophys. (1968) 125: 189-198.
24. Frederick, P. and Heath, R.L. Plant Physiol. (1974) (in press).
25. McNair-Scott, D.B. and Lesher, E.C. J. Bacteriol.) (1963) 85: 567-576.
26. Cleland, W.W. Biochem. (1964) 3: 480-482.
27. Mudd, J.B., McManus, T.T. and Ongun, A. "Second International Clean Air Congress" (H.M. Englund and W.T. Beery, Eds.) Academic Press, Inc., New York. (1971).

# 6

# Mechanisms of Ozone Injury to Plants

SAUL RICH and HARLEY TOMLINSON

Connecticut Agricultural Experiment Station, New Haven, Conn. 06504

Ozone and related oxidants are estimated to be responsible
for about 95% of the annual $130 million crop loss caused by air
pollutants in the United States. Reports have indicated that
ozone can seriously damage important crops such as spinach,
beans, petunias, citrus, tobacco, soybeans, and alfalfa, and
forest trees such as Eastern white pine and Ponderosa pine.
Susceptible cultivars of most of these plants develop severe
leaf injury when exposed to 2 to 5 pphm of ozone for 1 to 4 hr.
This level of ozone is very common in urban areas and so are
symptoms of ozone injury. However, such symptoms have also been
reported from plants growing in such rural states as Maine and
South Dakota. Even when no obvious injury can be seen, plants
exposed to low levels of ozone may not grow as well or yield as
much as plants growing in air free of ozone. On some plants,
e.g., spinach, the symptoms of ozone injury are quite distinc-
tive; consequently, plants like these are being used in some
places to detect and monitor air pollution.

In Connecticut, ozone injury was first seen on tobacco 20
years ago. This shade-grown crop used for cigar wrappers has an
annual cash value of about $22 million. In some years, ozone
damage caused a loss of up to $5 million. Plant breeders soon
produced cultivars both highly resistant to ozone and yet with
the necessary commercial qualities. With the adoption of these
cultivars, annual losses to the Connecticut tobacco crop from
ozone dropped to only a few thousand dollars, even though the
level of pollution did not recede.

This report includes a possible anatomical basis for the
flecking symptom and a summary of our search for the mechanisms
by which ozone injures plants.

## Anatomical Basis of Flecking

Flecking on the upper surface of leaves is a common symptom
of ozone injury on dicotyledonous plants. A single fleck is a
small line of dead tissue that appears white, yellow, or brown

against the green living tissue that surrounds it.

When the fleck is examined closely, the lesion can be associated with contiguous stomata in the upper surface. Often, the first visible symptom of ozone toxicity is the death of the palisade parenchyma cells that line the cavities directly beneath the upper stomata. In the case of beans (Phaseolus vulgaris L.) or tobacco (Nicotiana tabacum L.) and perhaps other plants, the upper stomata lie in patterns of arcs or circles, while the lower stomata are scattered randomly and regularly across the epidermis.

Adjacent upper stomata appear to be connected by air passages through the palisade parenchyma. These passages can be seen easily when a detached leaf is filled with water under pressure by injecting it through the petiole with a hypodermic syringe. The water is forced along the path of least resistance and fills the air spaces connecting the substomatal chambers of the upper stomata.

Apparently the palisade cells lining the substomatal cavities are the cells most sensitive to ozone in these leaves. As the cells are killed by ozone along the air passages connecting the substomatal chambers, the elongated fleck is formed. The random pattern of stomata in the lower surface, the different geometry of the air passages through the spongy parenchyma, and perhaps the greater resistance of spongy parenchyma cells to ozone make it unlikely that flecking would appear on the lower surface of these leaves.

Levels of ozone that do more than fleck the upper surfaces can cause irregular spots and blotches of dead tissue on the lower surface. Sufficient ozone to kill areas in the lower surface are usually enough to cause the entire area of the leaf to collapse, leading to the formation of "bifacial" lesions.

## Biochemical Studies

To determine how plants are injured, we studied changes in certain cellular constituents in plants exposed to ozone. The cells of these plants leak their contents and so it is probable that the initial damage is to cellular membranes. The normal functioning of these membranes depends on lipid constituents probably stabilized by sulfhydryl groups in associated proteins.

Sulfhydryl Groups. Sulfhydryl linkages are considered to be important to the structural integrity of membrane proteins.

In our first experiments (1) we subjected bean, spinach (Spinacia oleracea L.) and tobacco leaves to ozone at 1 ppm for 30 to 60 min. At this high concentration of ozone, the sulfhydryl content of the leaves was diminished 15 to 25% (Table I). There was little difference between the sulfhydryl content of ozone-resistant and ozone-susceptible tobacco either before or after ozonation.

Table I.  The effect of ozone on the concentration of
          sulfhydryl (SH) groups in leaves

| Time of analysis in relation to treatment[1] | Bean | Spinach | Tobacco | |
|---|---|---|---|---|
| | | | Ozone-resistant | Ozone-susceptible |
| | μmoles SH/g fresh tissue | | | |
| Immediately before | 1.20 | 1.65 | 0.67 | 0.74 |
| Immediately after | 1.10 | 1.50 | 0.60 | 0.67 |
| 30 Min after | 0.90 | 1.35 | -- | -- |
| 60 Min after | 0.90 | | 0.56 | 0.63 |

[1]Plants were treated with 1 ppm ozone for various times:  beans,
30 min; spinach, 45 min; tobacco, 60 min.

In another experiment (1) we treated ozone-resistant and
ozone-susceptible varieties of tobacco with toxic doses of
α-iodoacetic acid, and α-iodoacetamide, both sulfhydryl-binding
reagents.  The symptoms produced by both compounds were similar
to those produced by ozone.  The severity of the injury also
paralleled ozone resistance (Table II).  The degree of injury
caused by these two compounds also paralleled the ozone susceptibility of leaves of different ages on the same plant.  The upper-
most, youngest, leaves appear to be most resistant to both the
sulfhydryl-binding reagents and to ozone.

Table II.  Damage from sulfhydryl-binding reagents to the
           detached leaves of two tobacco varieties

| Leaf position[1] | % Leaf surface showing visible damage | |
|---|---|---|
| | Ozone-resistant variety | Ozone-susceptible variety |
| | α-iodoacetamide (10-2M) for 4 hr | |
| 1 | | 10 |
| 2 | | 20 |
| 3 | | 40 |
| 4 | 40 | 75 |
| | α-iodoacetic acid (10-3M) for 24 hr | |
| 4 | 50 | 85 |

[1]Position 1 is that of the youngest fully expanded leaf.
Position 3 and 4 are consecutively lower down the stem.

In later experiments (2), beans were subjected to a milder, longer exposure to ozone (25 pphm for 3 hr). This treatment did not diminish the sulfhydryl content appreciably, even though the ozonated leaves showed injury 18 hr later. However, we were able to detect newly produced disulfides (Table III). We concluded that ozonation changes proteins sufficiently to expose and oxidize additional sulfhydryl groups.

Table III.  The sulfhydryl content ($\mu$moles/g fresh wt) of opposite bean leaves, A and B, before and after exposure to ozone (25 pphm for 3 hr)

| Control[1] | | Ozonated[1] | | Ozonated, then dark[1] | |
|---|---|---|---|---|---|
| Leaf A | Leaf B | Leaf A | Leaf B | Leaf A[a] | Leaf B[a] |
| 1.45 | 1.45 | 1.35 | 1.50 | 1.20 | 1.35 |
| 1.45 | 1.50 | 1.35 | 1.60 | 1.35 | 1.50 |
| 1.50 | 1.55 | 1.50 | 1.60 | 1.40 | 1.50 |
| 1.60 | 1.55 | 1.70 | 1.90 | 1.45 | 1.60 |
| 1.70 | 1.70 | 1.75 | 1.95 | 1.50 | 1.70 |
| 1.75 | 1.75 | 1.75 | 1.95 | 1.50 | 1.80 |
| 1.75 | 1.80 | 1.80 | 1.95 | 1.60 | 1.90 |
| | | 1.85 | 2.00 | 1.65 | 1.85 |
| Md=0.014[x] | | Md=0.175[x] | | Md=1.875[y] | |

[1] Leaf A ground in polyvinylpyrrolidone medium without sodium sulfite.  Leaf B ground in same medium with sodium sulfite.  Control leaves not exposed to ozone or subjected to dark.  Dark period is 18 hr following exposure to ozone.

[x] Not significant.

[y] Significant beyond 1% level.

Lipid Metabolism.  Next we explored changes in lipid metabolism in leaves exposed to ozone.  Sterols and sterol derivatives were particularly interesting to us because they have been associated with membrane-containing fractions of leaves (3). Changes produced in these compounds may be early events in the toxicity of ozone to plant cells.

We first studied the changes in free sterols of leaves and chloroplasts of beans and spinach exposed to ozone (4).  We found (Table IV) that ozonated bean leaves and chloroplasts had 25% and 12% less free sterols respectively than bean leaves and chloro-

plasts that had not been ozonated. Ozonated spinach leaves and
chloroplasts had 44% and 37% less free sterols respectively than
their non-ozonated counterparts.


Table IV.  Changes in free sterol content of whole tissue and
           chloroplasts of bean and spinach leaves exposed to
                     ozone  (50 pphm for 1 hr)

| Plant part | Free sterol content[1] (μmoles/3 mg chlorophyll) | |
|---|---|---|
| | Control | Ozonated[x] |
| Bean | | |
|    Whole leaves | 0.83 | 0.62 |
|    Chloroplasts | 0.33 | 0.26 |
| Spinach | | |
|    Whole leaves | 1.16 | 0.65 |
|    Chloroplasts | 0.43 | 0.27 |

[1] Each value represents the mean of three experiments.  Each
experiment consisted of duplicate readings on leaf extracts of
two plants.

[x] Loss of free sterol significant beyond 1% level.


    In another experiment (Table V), the free sterol content of
ozonated chloroplasts from beans was found to be 32% less and the
content of sterol derivatives 37% more than that of non-ozonated
chloroplasts.  What happens to the free sterols (FS), sterol
glycosides (SG) and acetylated sterol glycosides (ASG) can be
seen in Table VI.  In these experiments (5) with whole leaves of
beans, FS in the ozonated leaves was 21% less, SG 32% more, and
ASG 41% more than in non-ozonated leaves.
    One of the interesting effects of ozone is the 56% increase
in the linolenic acid content of ASG from ozonated bean leaves
(5).  This led us to explore the source of the additional linolenic acid.  Ongun and Mudd (6) had reported that SG and ASG
normally formed at the expense of free sterols in non-ozonated
plants.  What happens in ozonated plants?
    Using bean leaf discs fed 1-[14]C-acetate, we showed that the
radioactively labeled diglyceride content of ozonated discs
became consistently less, and that this reduction was accompanied
by an increase in the radioactivity of ASG (unpublished).
    We proposed that this was caused by an increase in the rate
of lipid hydrolyses in ozonated discs.  To support this proposal,
we showed that fluorescein dilaurate was hydrolyzed 3 to 4 times

faster in ozonated discs than in non-ozonated discs (unpublished).


Table V.  Changes in free sterol and sterol derivative of
          chloroplasts in bean leaves exposed to ozone
          (50 pphm for 1 hr)

| Sterol Component | μ moles/3 mg chlorophyll[1] | | | | | |
|---|---|---|---|---|---|---|
| | Control | | | Ozonated | | |
| Free sterol | 0.38 | 0.39 | 0.35 | 0.24 | 0.25 | 0.26[x] |
| Sterol derivatives | 0.10 | 0.11 | 0.09 | 0.14 | 0.15 | 0.17[y] |

[1] Each value is the mean of duplicate readings made from 3 g of leaves.

[x] Significantly different beyond 1% level.

[y] Significantly different beyond 5% level.


Table VI.  Effect of ozonation (25 pphm for 2.5 - 3.0 hr) on
           free sterols (FS), sterol glycosides (SG), and
           acetylated sterol glycosides (ASG) in bean leaf tissue

| Sterol | Conc. (μmoles/10 Discs) | | |
|---|---|---|---|
| | Before $O_3$ | After $O_3$ | Diff. |
| FS | 0.93 | 0.73 | -0.20 |
| SG | 0.25 | 0.33 | +0.08 |
| ASG | 0.16 | 0.27 | +0.11 |


At this point, it is worth considering the importance of lipid peroxidation as a toxic mechanism in cells exposed to ozone. Scott and Lesher (7) proposed that ozone injures cell membranes by oxidizing unsaturated lipids. Goldstein and Balchum (8) reported that ozone reacts with unsaturated lipids to produce organic peroxides which, they suggested, poison cells. They used a thiobarbituric acid-malonyl dialdehyde (MDA) method to measure lipid peroxidation. Using this method, we could find no increase in MDA until after visible injury appeared on bean leaves (9). We concluded that lipid peroxidation may result from ozone injury to bean leaves rather than being the cause of injury.
    The objection has been raised that although the MDA test does

82 AIR POLLUTION EFFECTS ON PLANT GROWTH

detect lipid peroxidation, it can also be produced by the peroxidation of fatty acids. However, the increase in MDA only after visible injury on ozonated leaves allows us to conclude (a) that lipid peroxidation occurs after the initial toxic event, or (b) that lipid peroxidation does not occur at all. In either case, lipid peroxidation can be minimized as a cause of ozone injury to the plants that we studied.

Finally, we conclude that toxic reactions in plant cells injured by ozone probably take place in the following sequence: sulfhydryl oxidation, lipid hydrolysis, cellular leaking, lipid peroxidation, and then cellular collapse.

## Literature Cited

1. Tomlinson, H. and Rich, S. Phytopathology (1968) 58:808-810.
2. Tomlinson, H. and Rich, S. Phytopathology (1970) 60:1842-1843.
3. Grunwald, C. Plant Physiol. (1970) 45:663-666.
4. Tomlinson, H. and Rich, S. Phytopathology (1973) 63:903-906.
5. Tomlinson, H. and Rich, S. Phytopathology (1971) 61:1404-1405.
6. Ongun, A. and Mudd, J. B. Plant Physiol. (1970) 45:255-262.
7. Scott, D. B. M. and Lesher, E. C. J. Bacteriology (1963) 85:567-576.
8. Goldstein, B. D. and Balchum, O. J. Soc. Exp. Biol. Med. Proc. (1967) 126:356-358.
9. Tomlinson, H. and Rich, S. Phytopathology (1970) 60:1531-1532.

# Further Observation on the Effects of Ozone on the Ultrastructure of Leaf Tissue

WILLIAM W. THOMSON, JERRY NAGAHASHI, and KATHRYN PLATT

Biology Department, University of California, Riverside, Calif. 92502

## Introduction

In previous studies of ozone effects on the ultrastructure of leaf tissue, it was reported that the earliest changes observed involved the chloroplasts (1, 2) and an apparent swelling of the mitochondria (2). Thomson et al. (1) noted that the first changes in the chloroplast involved an increased granulation and electron density of the stroma with the subsequent appearance of crystalline arrays of fibrils in the stroma. In more recent studies with tobacco, Swanson et al. (2) did not observe the crystalline structures; however, they did observe early changes in the chloroplasts. These changes consisted of an increase in the density of the stroma, a decrease in chloroplast volume and a change in the configuration of the chloroplasts in that they were irregular in outline. Although these alterations occurred before any ultrastructural breakdown or disruption of cellular membranes was observed (1, 2), it has been suggested that they probably result from an ozone-induced molecular alteration, and resultant change in permeability of membranes (2, 3).

The present report elaborates the ozone-induced changes in the fine structure of bean leaf mesophyll cells with particular reference to both the crystalline bodies in the chloroplasts and early changes in membrane ultrastructure.

## Materials and Methods

Fourteen day old bean plants (Phaseolus vulgaris L., lot D415, Burpee Seed Company, Riverside, California) were used in these experiments. Plants were exposed to 0.4 ppm $O_3$ for one hr in a fumigation chamber with the procedures and conditions outlined previously (1, 2). Samples of the leaves were taken at the end of the fumigation period, and at one, two and four hr intervals after the fumigation period. Control samples for comparison were taken from untreated leaves. The samples were prepared for study with the electron microscope by fixation at

room temperature in 2.5% glutaraldehyde in 0.1M phosphate buffer
(pH 7.0) for one hr followed by one hr post-fixation in 1%
phosphate buffered osmium tetroxide.  The samples were dehydrated
in an acetone series and embedded in an epoxy resin (4).  Thin
sections were cut on a Porter-Blum MT-2 ultramicrotome, picked up
on uncoated copper grids and stained first with uranyl acetate
and then with lead citrate (5).  The sections were viewed with a
Philips EM 300 electron microscope.

## Observations

As in previous studies, mesophyll cells exhibited a range of
alterations from slight to severe in all samples taken after
fumigation with ozone.  However, the number of cells affected and
the extent of alterations observed were much less in samples
taken from leaves at the end of the fumigation period than in
the samples taken subsequently.  The number of cells exhibiting
changes, the extent of the alterations and the number of cells
with severe damage and disruption increased with time from the
end of the fumigation period.  We have observed characteristic
ultrastructural features which can be associated with mild,
moderate and severely damaged cells, and have interpreted these
as indicators representing a progressive pattern of changes and
damage induced by ozone.

As reported by Swanson, et al. (2) for tobacco, the first
apparent incipient changes induced by ozone consisted of an
alteration of the configuration of chloroplasts (Fig. 1).  The
chloroplasts were frequently irregular in outline with noticeable indentation, particularly in the surface adjacent to the
plasmalemma (Fig. 1, arrows).

In many of the chloroplasts exhibiting incipient changes,
clusters of ordered arrays of fibrils and crystalline bodies
were observed in the stroma (Fig. 2, 3F).  These fibrils and
crystalloids were frequently associated with the chloroplast
envelope and were often observed in conjunction with regions
where small but noticeable alterations of the ultrastructure of
the envelope were apparent.  These alterations consisted
primarily of an increase in the staining density of the envelope
membranes (Fig. 2, arrows) and a frequent accumulation of
electron-dense material between the two membranes of the envelope (Fig. 3, arrows).  Larger deposits of electron-dense material also occurred in association with the membranes in these
areas.  Another unusual observation was the frequent presence of
what appears to be a complex of several membranes including the
chloroplast envelope membranes associated with the indentation
of the chloroplast surface (Fig. 1, arrows) and the crystalloids
in the stroma (Fig. 3F).

Moderately damaged cells were characterized by the presence
of several, often exceedingly large, crystalloids in the chloroplast stroma (Fig. 4, 5).  The crystalline bodies occurred

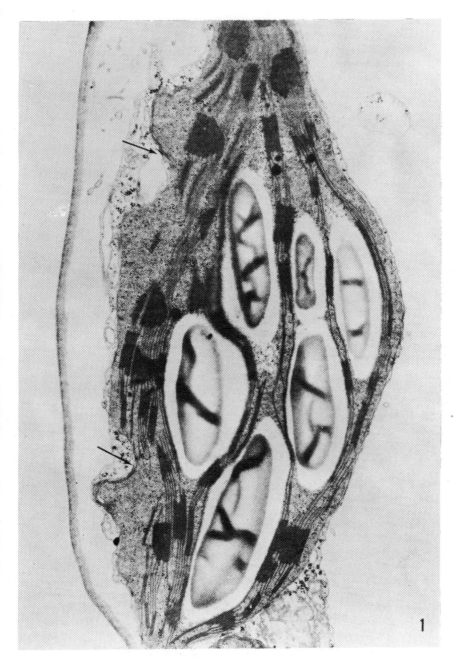

*Figure 1.  An irregular-shaped chloroplast which is characteristic of incipient, ozone-induced changes (x30,000).*

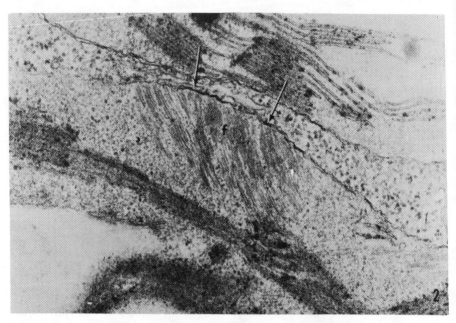

Figure 2.   A portion of a chloroplast exhibiting mild, ozone-induced alterations. Note the occurrence of stroma fibrils, f, in close association with regions of the envelope which have an increased density (see arrows) (x76,000).

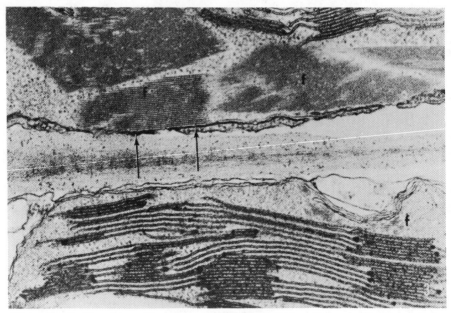

Figure 3.   Portion of two chloroplasts exhibiting mild, ozone-induced alterations. Note the close association of fibrils, f, with the envelope in regions where alterations such as an accumulation of electron dense material or indentations and membrane proliferations are apparent (x63,700).

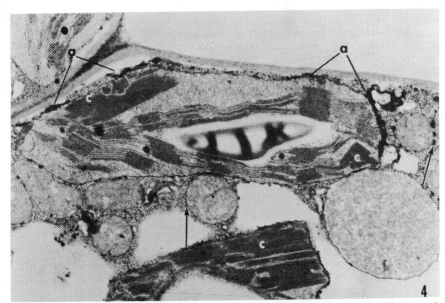

Figure 4. A portion of a cell exhibiting moderate ozone-induced changes. The crystalloids are generally distributed in the peripheral regions of the stroma and associated with the envelope. Note the accumulation of electron-dense material with the surface of the chloroplast, a, and the mitochondria, arrows (x27,700).

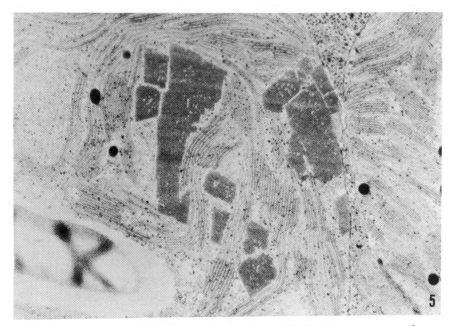

Figure 5. A portion of a chloroplast with several crystalloids. Note a presence of crystalloids in association with the envelope as well as around and between the grana (x51,000).

predominantly in the peripheral stromal regions adjacent to and associated with the chloroplast envelope (Fig. 4C); however, crystalloids and extensions of crystalloids around the grana into the more internal regions of the stroma were also observed (Fig. 5, 6, 7). The crystalloids were found to have two patterns of organization: generally they appeared to be composed of linear, electron-dense units, approximately 8.5nm wide separated by an electron-transparent band (Fig. 6); quite frequently, however, they had a "crosshatched" appearance (Fig. 7). As pointed out by deGreef and Verbelen (6), the two images probably result from sectioning in different planes. Defects, probably due to missing elements, were commonly observed in both images of the crystalloids (Fig. 5, 6, 7, 8).

As in the cells exhibiting mild or incipient damage, the crystalloid bodies in the moderately damaged cells were commonly associated with regions of the chloroplast envelope which were obviously altered. However, the changes and apparent damage to the envelope in these regions were more severe and extensive than in the mildly damaged cells. The changes in the envelope in these regions consisted of an increased staining density (Fig. 6, arrows), an increased accumulation of dense material between the two membranes, and an increase in size and amount of electron-dense material associated with these regions, (Fig. 4, 7, 8A). The envelope was often observed as a single entity (Fig. 7, 8S) and on occasions was not present at all (Fig. 7, arrow). In these moderately damaged cells, electron-dense accumulations also were often observed in association with the bounding envelope of the mitochondria (Fig. 4, arrows). Other membranes in these cells such as the tonoplast, microbody membrane, and plasmalemma did not appear damaged, although mild-to-severe plasmolysis was not uncommon.

In severely damaged cells the cell contents were collapsed into an aggregated mass (Fig. 9). Few cellular membranes were evident except for the internal membrane system of the chloroplasts (Fig. 9G). The crystalline arrays were distributed throughout the aggregated mass (Fig. 9C) and accumulations of electron-dense material occurred generally in association with the periphery of the aggregated mass (Fig. 9, arrows).

Discussion

The most obvious ozone-induced change in the mild- and moderately-damaged mesophyll cells was the development of small to large crystalloids in the stroma of the chloroplasts. Similar crystalloids have been described in previous studies on oxidant-induced damage to bean leaves (7). They have also been described in chloroplasts of bean and other plant leaves under various conditions of water stress (6, 8, 9, 10, 11, 12), as well as in chloroplasts of leaves treated with other phytotoxic air pollutants (13). It has also been suggested that in the

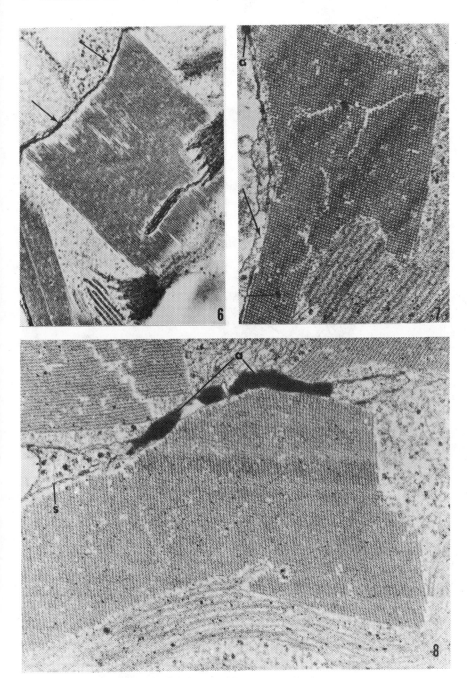

*Figures 6, 7, and 8. Crystalloids in chloroplasts of moderately damaged cells. Note the close association of the crystalloids with regions of the envelope where noticeable alterations have occurred. Magnification 64,000; 97,000; 102,000 respectively.*

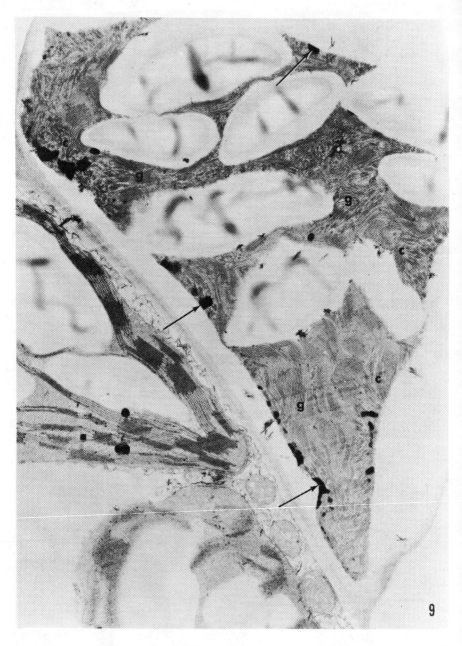

Figure 9.   A portion of a cell exhibiting severe damage, upper right, adjacent to a cell
with no apparent damage, lower left (x25,600).

ozone treated material the change in configuration, reduction in volume, and formation of the crystalloids in chloroplasts is related to the loss of water from the chloroplasts resulting in dehydration of the stroma (2, 10, 12). The most probable reason for water loss from chloroplasts would seem to be an ozone-induced increase in permeability of the chloroplast envelope (2). Recent experimental support for this possibility has been presented by Nobel and Wang (14) who found that ozone increased the permeability of isolated chloroplasts.

The concomitant appearance of localized alterations in the chloroplast envelope with the crystalloids in the stroma as well as the coincidence in the occurrence of these crystalloids adjacent to the altered membrane regions suggests that a direct relationship exists. One explanation that can be offered is that the changes in the envelope, i. e., increased staining density of the membranes, accumulation of electron-dense material between the two membranes, and the accumulation of electron-dense materials associated with the membranes represent degradative alterations of the membranes and are probably localized sites of ozone-induced damage to the envelope. The probable altered permeability of these regions would result in the dehydration of the adjacent stroma and the early formation of crystalloids adjacent to these damaged regions of the membranes.

The increase in the extent and intensity of the envelope alterations in the moderately-damaged cells tends to support the conclusion that these changes are degradative and are sites of ozone-induced damage. The similarity of the electron-dense accumulation in association with mitochondrial membranes to those associated with the chloroplast envelope also suggests that such accumulations are characteristic of degradative changes in membranes. In previous studies of ozone-induced damage to mesophyll cells, similar accumulations were observed in moderate- to severely-damaged cells (1); the accumulations in severely damaged cells probably represent aggregates of membrane material released during the breakdown of the membranes.

One of the major suggestions about the primary site of ozone reaction with cells (15) is that it damages cell membranes. The results of this study support this contention: in the moderate- to severely-damaged cells, except for the internal membrane system of chloroplasts, membrane degradation and loss of cell compartmentation is evident. In mild- and incipiently-damaged cells, however, membrane alterations are apparently confined to localized regions of the chloroplast envelope. These observations suggest that the early effects of ozone on membranes are limited to these regions.

There are several lines of research which indicate that ozone can react with proteins, amino acids and olefinic groups of membrane lipids (16, 17). Regarding unsaturated fatty acids, Swanson et al. (2) concluded that these olefinic groups were unlikely to be the primary site of ozone action since the relative amounts of fatty acids extracted from ozone-treated and untreated leaf tissue were almost equivalent. However, they pointed out that their procedures would probably not detect small changes in membrane lipids such as might occur in the localized membranes of the envelope observed in the present studies.

There is another reason to suspect that the primary reaction of ozone in these localized regions of the envelope is not with the unsaturated fatty acids, but rather with other moieties or parameters of the membranes. The increased staining density of the membranes and the electron-dense accumulations in these regions probably represent components released from the membrane which, upon release, are available for reaction with osmium. The most likely identity of these components are the membrane lipids. We suggest, therefore, that the primary reaction of ozone with the envelope membranes involves the disorganization of the membranes via constituents other than the lipids. A similar possibility has been developed by Mudd et al. (17) who found that ozone does not react readily with lecithin when in a bilayer configuration. In addition, Swanson et al. (2) reported that the electron microscopic image of most cellular membranes is not affected by prior treatment with ozone. The rationale in their discussions is that the architecture of the membrane restricts the ability of ozone to react with the olefinic groups.

## Literature Cited

1. Thomson, W. W. Dugger, W. M. Jr. and Palmer, R. L. Dan. J. Bot. (1966) 44: 16-77-1682.
2. Swanson, E. S. Thomson, W. W. and Mudd, J. B. Can. J. Bot. (1973) 51:1213-1219.
3. Thomson, W. W. and Swanson, E. S. 30th Ann. Proc. Electron Microscopy Soc. Am. (1972) 360-361.
4. Spurr, A. R. J. Ultrastruct. Res. (1969) 26:31-43.
5. Reynolds, E. S. J. Cell Biol. (1963) 17:208-212.
6. De Greef, J. A. and Verbelen, J. P. Ann. Bot. (1973) 37:593-596.
7. Thomson, W. W., Dugger, W. M. Jr. and Palmer, R. L. Bot. Gaz (1965) 126:66-72.
8. Perner, E. Port. Acta Biol. (1962) A6:359-372.
9. Perner, E. Naturwissenshaften (1963) 50:134-135.
10. Shumway, L. K., Weier, T. E. and, Stocking, C. R. Planta (1967) 76:182-189.
11. Wrischer, M. Planta (1967) 75:309-318.
12. Wrischer, M. Protoplasma (1973) 77:141-150.
13. Dolzman, P. and, Ullrich, H. Z. Pflanzenphysiol (1966) 55:165-180.
14. Nobel, P. S. and, Wang, C-T. Arch. Biochem. Biophys. (1973) 157:388-394.
15. Rich, S. Ann. Rev. Phytopath. (1964) 2:253-266.
16. Mudd, J. B., Leavitt, R., Ongun, A. and McManus, T. T. Atmos. Environ. (1969) 3:669-682.
17. Mudd, J. B., McManus, T. T., Ongun, A. and, McCullogh, T. E. Plant Physiol. (1971) 48:335-339.

# Phenols, Ozone, and Their Involvement in Pigmentation and Physiology of Plant Injury

ROBERT K. HOWELL

Air Pollution Laboratory, Agricultural Environmental Quality Institute,
Northeastern Region, Agricultural Research Service, U.S. Department
of Agriculture, Beltsville, Md. 20705

Abstract

　　In plant cells, phenols and derivatives are located in chloroplasts and in vacuoles; enzymes that oxidize phenols are also
located in chloroplasts and in cytoplasm but are maintained in
separate compartments by membranes. Ozone impairs the integrity
of cell membranes and thus permits oxidative enzymes to oxidize
phenols to their respective quinones. o-Quinones have an $E_o'$ of
+1.9V and will polymerize with amino acids, amines and sulfhydryl
groups of proteins to form low molecular weight reddish-brown
pigments in leaves of ozone-treated plants. This involvement of
phenols appears to be the cause of the visible necrotic lesions
on injured leaves. Chemically, the pigments or polymers resemble
those formed in tobacco leaves during the curing processes.
Hydrolzates of the polymers contain several amino acids, metals
and a phenol. The polymers are lignin- or tannin-like and as
such detract from the esthetic and probably nutritional value of
foliage from important food and feed crops. Concentrations of
caffeoyl derivatives, caffeic and chlorogenic acids are increased in ozone-damaged tissues. Both o-diphenols increase $O_2$
consumption and reduce $CO_2$ fixation. Therefore, plant growth and
quality could be reduced by ozone's (a) impairing membrane
integrity which would promote cell degradation through reduction
in synthesis of products of primary metabolism and by (b) increasing products of secondary metabolism.

Introduction

　　One factor contributing to crop losses caused by ozone is the
development of abnormal pigments which are viewed as discolored
areas on foliage of ozone-sensitive cultivars. Such discoloration
is esthetically unacceptable on leafy food crops and probably
causes a reduction in the nutritional value of all foliar components used for food and feed. Biochemical and physiological
reactions responsible for ozone-induced discolorations have not

been explored to the same extent as ozone-induced physiological reactions associated with primary plant metabolism. Therefore, any discussion of the physiology of abnormal pigmentation resulting from ozone assaults has to consider the limited information available and relate the findings from other investigations involving plant stresses to findings from ozone-induced changes in plant metabolism.

Phenols and enzymes of the phenolase complex, o-diphenol: $O_2$ oxidoreductase (E.C. 1.10.3.1.) and peroxidase (E.C. 1.11.1.7.) contribute to secondary pigment formation during curing processes (1) and during stresses to plants caused by ozone (2,3), diseases, chemicals, physical wounding and adverse temperatures, moisture and nutrient levels (4, 5). The resulting pigments contribute to abnormal foliar and food coloration (6) and flavors (6, 7) of plants and plant products.

Ozone stimulates the synthesis of anthocyanins (2), caffeoyl derivatives (8), and total phenols (9) in plant foliage. It induces necrotic lesions and localized discolored areas of yellow, red or brown on foliage of plants of many species. The size and number of lesions and the intensity of color are related to plant cultivar, maturity of foliage, cultural conditions, concentration of ozone and duration of time after exposure.

Why phenolic constituents increase in ozone-injured tissues or what they contribute to plant injury is not clear. To support the concept that there is a relationship between plant phenolic concentrations and ozone injury, four approaches are presented: (a) array of plant species that produce pigment in response to ozone treatment; (b) timing of pigment development after ozone treatment; (c) plant resistance to ozone and absence of pigmentation; and (d) correlation between total phenols in foliage and relative sensitivity of peanut cultivars to ozone.

Species That Produce Pigments in Response to Ozone. Ozone causes pigment changes in many plant species. Chlorophyll degradation occurs and chlorosis results. In one plant, Rumex crispus L., chlorophyll concentrations did not decrease but anthocyanin synthesis was increased after exposure to ozone (2). In other plants, reddish-brown pigments are formed, but there is a decrease in chlorophyll content simultaneous with their formation (3). The reddish-brown pigments can be easily visualized and located in injured leaves after the remaining chlorophyll and other ethanol-soluble pigments are washed from leaves. Certain cultivars of the following plants typically exhibit polymerized pigments after ozone treatments: green beans, sunflower, potato, white and red clover, spinach, safflower, curly dock, sweet clover, sweet and grain sorghum, lima beans, chrysanthemum, sycamore, dogwood, peanuts, alfalfa, cotton, radish, buckwheat and soybeans. Certain cultivars of green bean and alfalfa are resistant to ozone and fail to produce the polymerized pigment following exposures to ozone concentrations that injure sensitive cultivars.

Time of Pigment Formation After Ozone Treatments. In general, pigmented lesions in green leaves of test plants are not visible for approximately 20 hr or more after ozone treatments of 10 pphm or less for 2-8 hr, depending upon cultivars used. If, however, alcohol-soluble pigments are removed from similar ozone-treated leaves, the oxidized pigments are easily visible within 2 hr after fumigations. Pigment formation is evident in Kent soybean leaves (Fig. 1), within 2 hr after ozone treatment, 10 pphm for 6 hr, and spreads to nearly the entire leaf by 4 hr after treatment. Little more pigment was detectable during the next 20 hr.

Resistance to Ozone and Failure to Produce Abnormal Pigments. Tempo green beans and alfalfa clones 11-1-1L are very sensitive to ozone; Greenpod 407 and alfalfa clone 2-1-2H are resistant to ozone. If the sensitive and resistant clones of the two species are cultured in ambient concentrations of ozone, the leaves of resistant plants do not have any visible lesions but those of the susceptible plants have extensive injury. After removing the alcohol-soluble pigments from all leaves, the leaves from susceptible plants are significantly more pigmented than those from resistant plants, whose leaves are essentially void of polymerized pigments.

Correlation Between Total Phenols Expressed as Percent Caffeic Acid Equivalents and Ozone Injury. Twelve cultivars of peanuts ranging from low to high ozone sensitivity were examined. Foliage of each of the cultivars was evaluated for content of total phenols before and after ozone treatment. Because of the relatively high concentrations of caffeic acid in peanut foliage, values are expressed as percent caffeic equivalents in the total phenol assay. Experimental results used in establishing the correlation between caffeic acid and ozone injury are shown in Table I.

## Discussion

It is a common phenomenon for phenolic moieties to accumulate in plant tissues after exposures to stresses. Concentrations of phenols are increased in ozone-treated plant tissues (3, 9). Reddish-brown polymerized pigments also develop in a diversity of plants injured by ozone. This response appears to be a general reaction and not a specific result of ozone injury (5). Similar pigments have been identified in plant tissues during senescence and by curing processes (1).

The products of oxidation are easily visualized shortly after ozone treatments but no claim should be made that phenolics or the oxidative enzymes, polyphenol oxidase, phenolase or peroxidase are the primary sites of ozone injury.

The involvement of phenols and enzymes of the phenolase complex appears to be secondary to the induction of necrosis.  The
induction must involve a modification of membrane structure which
leads to altered membrane permeability and loss of cell compart-
mentalization.  If this occurs, regulation of cellular metabolism
is lost, enzymes are activated, and these and their substrates
that are normally separated by membranes would react together.
The resulting oxidative activities would then lead to the
formation of pigments or lesions.

Table I.  Index of ozone injury to foliage of 12 peanut cultivars
and total plant phenol expresses as % caffeic acid
equivalents in assay

| Peanut cultivar | Injury score | Total plant phenol expresses as % caffeic acid equivalents in assay | |
| | | Before ozone treatment | After ozone treatment |
| --- | --- | --- | --- |
| PI268661 | 96 | 0.042 | 0.108 |
| Wilco | 66 | 0.038 | 0.073 |
| Florspan | 69 | 0.034 | 0.065 |
| NC4X | 60 | 0.033 | 0.055 |
| SER 56-15 | 70 | 0.039 | 0.080 |
| Spancross | 67 | 0.040 | 0.093 |
| NC 4 | 65 | 0.035 | 0.066 |
| Tennessee Red | 69 | 0.031 | 0.070 |
| Florunner | 56 | 0.038 | 0.053 |
| Tifspan | 70 | 0.044 | 0.098 |
| Virginia 61R | 70 | 0.030 | 0.065 |
| Florigiant | 61 | 0.028 | 0.068 |
| | Mean | 0.036 | |
| | $r = 98.6$ | | |

After loss of membrane integrity in a few susceptible cells,
chain reactions involving both intra- and extracellular components could commence.  The possibility of this type of overall
response is made more plausible when we consider the work

describing leaf age and ozone sensitivity (10, 11). In both Poa annua L. and Nicotiana glutinosa L., leaves show several gradations in cell maturation. Only specific areas on the leaves of both species were sensitive to ozone. The authors concluded that ozone sensitivity is a function of cellular development and maturity. The same conclusion was reported by Ting and Dugger (12). This information indicates that an intact leaf or plant is a dynamic system complete with cells or leaves of different maturities and, therefore, of different abilities to resist ozone uptake and subsequent biochemical changes. Generally, phenol metabolism is considered a secondary physiological process compared to photosynthesis. Therefore, the relative timing of an ozone effective contact on the two processes in a dynamic system (such as a leaf) may be only a matter of seconds and makes a discussion of what the primary or secondary site of ozone attack in an intact system pointless. A leaf is composed of many cells, each attached to an adjacent one. During leaf expansion, some cells are mature, others are enlarging while others are still dividing; therefore, in a given leaf, there should be some cells at a physiological age that would favor anabolism and photosynthesis, and some that would favor catabolism, or phenol polymerization.

In any discussion where an attempt is made to identify the primary or secondary sites of ozone reaction, one has to consider the leaf as having many cells, each of which under normal conditions would provide protection for subcellular organelles. As cells age, products of primary metabolism, such as proteins, carbohydrates and lipids are quantitatively reduced (13) and the stability of membranes diminished, thus allowing the biochemical reactions and products of secondary metabolism to dominate. In a tissue with cells of different ages, ozone could be affecting many metabolic reactions involving these components: (a) glucose (14), (b) protein (15), (c) lipid (16), and (d) phenols (8) at the same time. So the first site of ozone degradation is most probably in membranes.

Ozone changes plant cell membrane (17, 18) and chloroplast membrane permeability (19). Loss of membrane integrity is also caused by plant pathogens (20), drought (21), herbicides (22) and frost (21).

Phenols are present in chloroplasts (23, 24) and in vacuoles (25) of plant cells. The enzyme polyphenol oxidase and other enzymes of the phenolase complex are bound to the chloroplast lamellae or stroma (26, 27) and in the cytoplasm (26). Although the enzyme and substrate are in close proximity, they fail to react until a physical disruption or natural aging process alters membrane permeability and permits them to react. o-Phenols are rapidly oxidized to o-quinones in the presence of oxygen. Ozone could contribute to quinone formation in two ways: by (a) destroying membrane integrity to permit substrate and enzyme to react, and (b) providing molecular oxygen for phenol oxidation.

Ozone does affect photosynthesis as measured by $CO_2$ fixation
(28, 29). Since loss of membrane integrity and oxidized pigment
formation occur within 2 hr after ozone treatment, there is
reason to suspect that either endogenous o-phenols or their oxidized products, formed after membrane disruption, may explain the
relationship of ozone, phenols and reduced $CO_2$ fixation, and $O_2$
evolution. Chloroplasts of plant species differ in their endogenous quantities of o-phenols (23) and polyphenol oxidase enzymes
(26). Baldry et al. (23) have compared the photosynthetic
abilities of chloroplasts isolated from sugar cane and spinach by
measuring $CO_2$ fixation and $O_2$ evolution. $CO_2$ fixation and $O_2$
evolution were both significantly more suppressed by chloroplasts
from sugar cane than by those from spinach. If spinach chloroplasts were suspended in extracts from sugar cane chloroplasts
which contained mainly o-diphenols, $CO_2$-dependent $O_2$ evolution
was reduced by 100% of the control and photofixation of $CO_2$ was
reduced to only 3% of that observed for nontreated spinach chloroplasts. The predominant o-diphenols were identified as chlorogenic and caffeic acids. Both o-diphenols at concentrations of
$2 \times 10^{-3}$ M suppressed $CO_2$ fixation and $O_2$ evolution; however, of
the two, caffeic acid was the more effective. An exogenous
electron acceptor added to the chloroplast preparation negated
the effect of the o-diphenol and $O_2$ evolution occurred. Polyphenol oxidase is bound to chloroplast lamellae, and Tolbert (26)
showed that the ability of chloroplast preparations to oxidize
dihydrophenylalanine differs depending upon plant species, intensity and quality of light, and aging of chloroplasts. Chloroplasts from sugar beet, Swiss chard, and spinach had low Km values
initially; but the values increased to a maximum of
$2-6$ mmole $\cdot$ mg$^{-1}$ chlorophyll $\cdot$ hr$^{-1}$ after aging for 1-5 days.
Chloroplasts isolated from alfalfa, wheat, oats, peas and sugar
cane leaves had maximum Km values of $11 - 120$ µmoles $\cdot$ mg$^{-1}$
chlorophyll $\cdot$ hr$^{-1}$ and did not require aging for maximum oxidation
rates. Thus, plant enzyme properties differ and such differences
have to be known before ozone and its effect on biochemical
systems can be adequately interpreted. Suppression of photosynthesis could be due to o-diphenols and o-quinones: (a) acting
as electron donors or acceptors which uncouple the normal pathway of photosynthetic electron flow or (b) changing the redox
($E_o'$) potentials of intermediates of the electron transport chain.
Partial photoreduction of the o-diquinones also produces highly
reactive and inhibitory semi-quinones (23).

Mason (30) and Pierpoint (31) have described the involvement
of o-diphenols in plants and how they contribute to abnormal
plant pigmentation. o-Diphenols are oxidized to o-quinones by
enzymes of the phenolase complex (o-diphenol: $O_2$ oxidoreductase,
E.C. 1.10.3.1) and by peroxidase (E.C. 1.11.1.7). o-Quinones
react with amino acids, proteins, amines and thiol groups of
proteins to polymerize and from reddish-brown pigments. Concentrations of caffeic acid are doubled in both bean (8) and peanut

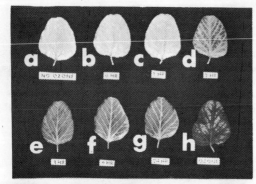

*Figure 1–1. Kent soybean leaves: a. No ozone but cleared in 95% ethanol. b–g. Treated with ozone 10 pphm 6 hr, removed from plants 0, 1, 2, 4, 6, or 24 hrs after treatment and cleared. h. Ozone injury 24 hrs after treatment but not cleared.*

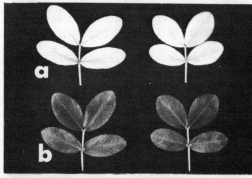

*Figure 1–2. Tifspan peanut leaves: a. No ozone and cleared. b. Treated with ozone and cleared.*

*Figure 1–3. King's Ransom chrysanthemum leaves: a. Three on the left—treated with ozone and cleared. b. Two on the right—no ozone and cleared.*

*Figure 1–4. Alfalfa leaves exposed to ambient concentrations of ozone: a. From a resistant clone and shows no injury or pigment accumulation. b. From a susceptible plant and shows both leaf injury and pigment accumulation.*

*Figures 1–5 and 1–6. Green bean leaves exposed to ambient concentrations of ozone: a. Resistant and shows no pigment. b. Sensitive and shows that pigment is accumulated.*

*Figure 1–7.    Greenpod 407 leaf with bean rust pustules.*

*Figure 1–8. Spinach leaves: a. Treated with ozone and cleared to expose pigment. b. No ozone but cleared.*

leaves after ozone exposures. Ozone exposures also increased
peroxidase activity in bean leaves (32, 33). Increased concentrations of total phenols occur in tobacco leaves as a result of
oxidant exposures (9). The reddish-brown polymers which form in
ozone-injured leaves are presumably products that form from reactions of ozone ($E_0^1$ +2.07V) with phenolic moieties which are
oxidized to o-quinones ($E_0^1$ +1.90V) and which then polymerize with
amino acids, proteins, amines or sulfhydryl compounds.

Plant susceptibility to ozone as determined by visible injury may be very closely related to quantities of o-diphenols
associated with the chloroplasts and specific requirements for
activation of polyphenol oxidase enzymes. There is a significant
correlation between ozone injury and concentrations of total
phenols expressed as percent caffeic acid equivalents in peanut
cultivars. This concept is not intended to underestimate the
importance of membranes that separate phenols and enzymes. Perhaps future research will demonstrate that membranes of resistant
alfalfa, green bean and other species differ both qualitatively
and quantitatively from those of susceptible plants of these
species.

Phenols are responsible for plant resistance to certain
pathogens (5, 20, 34). Because of increased interest in biological pest control, selection of plants with high concentrations of phenols may provide protection from plant pests, but on
the basis of our knowledge of the distribution of oxidant pollutants and their effects on plant metabolism such a hypothesis
would be undesirable. Ozone is widespread and does contribute to
the oxidation of phenols in foliage of many plant species.
Resulting polymers are lignin- or tannin-like, and detract from
both the esthetic and probably the nutritional values of foliage
from such crops as alfalfa, clover and spinach. Products of
oxidation in Tempo bean leaves provided protection again against
a natural bean rust (Uromyces phaseoli (Pers.) Wint.) infection.
No fungal colonies were detected on injured Tempo bean leaves.
The cultivar Green Pod 407 that did not produce the polymer
(Fig. 1) was well inoculated. The bean rust pathogen is an
obligate parasite; therefore, it is difficult to determine if an
essential growth factor was immobilized and not available to the
rust organism or whether the pathogen was inhibited by phenols
or o-quinones.

Phenolic compounds naturally occurring in plants have induced many physiological responses that duplicate those reported
for ozone and/or peroxyacetylnitrate (PAN). Chlorogenic acid is
a competitive inhibitor of IAA-oxidase (35) and plant growth is
adversely affected by increased concentrations of auxins (36).
Concentrations of chlorogenic acid are increased in tobacco
tissue exposed to ozone (9). Phenols inhibit ATP synthesis (37),
oxidative phosphorylation (37) and SH⁻enzyme activity (27); they
increase respiration (38), reduce $CO_2$ fixation (22), modify both
membrane permeability (40) and oxidation rate of reduced NADH

(39), and form adducts that increase oxidation rates of amino acids (41). Dugger and Ting relate in their recent review (42) that ozone and/or PAN elicit the same physiological responses in suppressed growth resulting from ozone treatments could result from interactions of ozone, membrane components and phenols, and specific enzymes of the phenolase complex.

## Conclusions

Phenols, naturally found in plants modify pigmentation, quality and growth of those plants. Naturally occurring plant phenols, such as derivatives of caffeic or chlorogenic acid, elicit many physiological responses in plant assay systems. When similar assay systems are treated with ozone, responses similar to those stimulated by phenols are observed. Therefore, ozone is suspected of interacting with cellular membranes which permit phenoloxidase and phenols to react. As a result of such reactions, ozone maximizes the effects on pigmentation, and perhaps on quality and growth of plant parts. Crops developed for certain areas of the country, especially those subject to high oxidant concentrations, should be selected on the basis of either low phenol content or phenolase activity.

Acknowledgment. Grateful appreciation is expressed to Diane F. Kremer for her technical assistance. This work was supported in part under Contract EPA-IAG-D4-0479, Environmental Protection Agency, National Ecological Research Laboratory, National Environmental Research Center, Corvallis, Oregon 97330.

## Literature Cited

1.  Forsyth, W. G. C.  Ann. Rev. Plant Physiol. (1964)
    15:443-450.
2.  Koukol, J. and Dugger, W. M., Jr.  Plant Physiol. (1967)
    42:1023-1024.
3.  Howell, R. K. and Kremer, D. F.  J. Environ. Qual. (1973)
    2:434-438.
4.  Craft, C. C. and Audia, W. V.  Bot. Gaz. (1962) 123:211-214.
5.  Farkas, G. L. and Kiraly, Z.  Phytopathol. A. (1962)
    44:105-150.
6.  Swain, T.  In T. A. Geissman (ed.) "The chemistry of
    flavonoid compounds." Macmillan and Co., New York.
7.  Singleton, V. L.  In C. O. Chichester (ed.)  "Chemistry of
    plant pigments." Academic Press, New York. (1972).
8.  Howell, R. K.  Phytopathology (1970) 60:1626-1629.
9.  Menser, H. A. and Chaplin, J. F.  Tobacco Sci. (1969)
    73:73-74.
10. Bobrov, R. A.  Amer. J. Bot. (1955) 42:467-470.
11. Glater, R. B., Solberg, R. A. and Scott, F. M.  Amer. J.
    Bot. (1962) 49:954-970.
12. Ting, I. P. and Dugger, W. M., Jr.  J. Air. Poll. Control
    Assoc. (1968) 18:810-813.
13. Woolhouse, H. W.  In H. W. Woolhouse (ed.) "Twenty-first
    Symposium Society Experimental Biology on Aspects of the
    Biology of Ageing." pp. 179-213. Academic Press, New York
    (1967).
14. Dugger, W. M., Jr., Koukol, J. and Palmer, R. L.  J. Air
    Poll. Control Assoc. (1966) 16:467-471.
15. Chang, C. W.  Phytochemistry (1971) 10:2863-2868.
16. Tomlinson, H. and Rich, S.  Phytopathology (1969)
    59:1284-1286.
17. Evans, L. S. and Ting, I. P.  Amer. J. Bot. (1973)
    60:155-162.
18. Perchorowicz, J. T. and Ting, I. P.  Amer. J. Bot. (1974)
    In press.
19. Nobel, P. S. and Wang, C. T.  Arch. Biochim. Biophys. (1973)
    157:388-391.
20. Kosuge, T.  Ann. Rev. Phytopath. (1969) 7:195-230.
21. Santarious, K. A.  Planta (1973) 113:105-109.
22. Ashton, F. M. and Crafts, A. S.  "Mode of action of
    herbicides." pp. 1-394.  John Wiley & Sons, New York (1973).
23. Baldry, C. W., Bucke, C., Coombs, J. and Gross, D.  Planta
    (1970) 94:107-113.
24. Monties, B.  Bull. Soc. Franc. Physiol. Veg. (1969)
    15:29-45.
25. Pridham, J. B.  "Phenolics in plants in health and disease."
    pp. 9-15.  Pergamon Press, Oxford, London (1960).
26. Tolbert, N. E.  Plant Physiol. (1973) 51:234-240.

27. Butt, V. S.  Hoppe-Seyler's Z. Physiol. Chem. (1972)
    353:131.
28. Hill, A. C. and Littlefield, N.  Environ. Sci. Tech. (1969)
    3:52-56.
29. Bennett, J. H. and Hill, A. C.  J. Environ. Qual. (1973)
    2:526-530.
30. Mason, H. S.  In F. F. Nord (ed.) "Advances of Enzymology."
    pp. 105-184.  Interscience Publishers, New York (1953).
31. Pierpoint, W. S.  Biochem. J. (1969) 112:609-617.
32. Dass, H. S. and Weaver, G. M.  Can. J. Plant Sci. (1969)
    48:569-574.
33. Curtis, C. R. and Howell, R. K.  Phytopathology (1971)
    61:1306-1307.
34. Kuc. Joseph.  Ann. Rev. Microbiol. (1966) 20:337-370.
35. Morre, D. J. and Bonner, J.  Physiol. Plant (1965)
    18:635-649.
36. Stenlid, Göran.  Phytochemistry (1970) 9:2251-2256.
37. Dedonder, A. and Van Sumere, C. F.  Z. Pflanzenphysiol.
    (1971) 61:70-80.
38. Gamborg, O. L., Wetter, L. R. and Neish, A. C.  Can J.
    Biochem. Physiol. (1961) 39:1113-1124.
39. Keck, R. W. and Hodges, T. K.  Phytopathology (1973)
    63:226-228.
40. Hess, Earl H.  Ref. I. Manometric Studies.  Arch. Biochem.
    Biophys. (1958) 74:198-208.
41. Dugger, W. J., Jr. and Ting, I. P.  Ann. Rev. Plant Physiol.
    (1970) 21:215-234.

9

# The Impact of Ozone on the Bioenergetics of Plant Systems

EVA J. PELL

Department of Plant Pathology, Pennsylvania State University,
University Park, Penn. 16802

Abstract

Ozone has been shown to initiate many physiological and biochemical changes in sensitive plant species. Decreases in photosynthesis and increases and decreases in respiration have
occurred in response to ozonation. The bioenergetic status of
mitochondria and chloroplasts is disturbed by ozone. Decreases
in oxidative- and photo- phosphorylation have been reported as
have increases in adenosine triphosphate and total adenylate
content of plant tissue. The variable physiological responses
appear to be related to the stage of symptom development at the
time of analysis and to the mode of ozone exposure, viz. in vivo
and in vitro.

In recent years there has been increased interest in
elucidating the biochemical and physiological responses of plants
to air pollutants. Ozone has received particular scrutiny
because of its importance in photochemical smog. The effects of
ozone on the spectrum of physiological and biochemical systems
has been analyzed in plant systems both in vivo and in vitro to
try to determine the mode of action of the gas at both the
cellular and subcellular level (1). The data obtained from this
research has allowed the characterization of ozone damage at
various stages of symptom development and at various cellular
levels. We now know that ozone has the potential to damage membrane systems, alter enzyme function and modify corresponding
metabolic pathways (1). However, because of ozone's rapid reaction rate, the primary site of its action is still open to
speculation. While the primary effects of ozone on the cell will
lead to the altered plant growth ultimately observed, the
connection may not be direct. Therefore, it would be appropriate
to consider the response of a regulatory system to ozone exposure.
Regardless of whether the response to ozone is primary or
secondary, the effects on the cell and entire plant system may be

permanent.

The study of bioenergetics deals with energy formation and utilization in living systems. In most living cells energy is found in the form of adenosine triphosphate (ATP). Many biochemists (2, 3) believe that the cell is regulated, at least in part, by the energy charge, i.e.

$$\text{energy charge} = \frac{[\text{ATP}] + \frac{1}{2}[\text{ADP}]}{[\text{ATP}] + [\text{ADP}] + [\text{AMP}]}$$

Imbalances in energy charge within a cell may be due to reduced availability and/or increased demand for ATP or its precursors. Reduced energy availability may result from impairment of the energy-generating unit and/or from dislocation of the high energy molecule from the site requiring ATP.

In the plant cell there are two sites of ATP generation: the chloroplasts and the mitochondria. Adenosine triphosphate is generated during the process of oxygen utilization, viz. during respiration and photosynthesis which results in oxygen evolution. Since ozone is an allotrope of oxygen, it was logical to examine the effects of the air pollutant on the uptake and evolution of oxygen by the mitochondria and chloroplasts, respectively, and to analyze these organelles for corresponding effects on phosphorylation. In the following pages I should like to discuss research which has been conducted to determine the effect of ozone on both ATP concentrations and the phosphorylating abilities of plant species; then I will correlate this research with studies defining the photosynthetic and respiratory behavior of plant systems.

Respiration and photosynthesis were among the first responses to ozone examined (4-9). Respiration has decreased under some circumstances (6, 7), but in general it has been shown to increase. These increases have been correlated with the appearance of visible symptoms (8). Since the generation of ATP is so closely associated with electron transport and oxygen uptake, the effects of ozone on oxidative phosphorylation were examined. Intact tobacco plants were exposed to 0.60-0.70 µl/l ozone for 1 hr; mitochondria isolated from visibly injured tissue demonstrated an inhibition of oxidative phosphorylation in conjunction with an increase in respiration (6). However, when detached tobacco leaves were fumigated with 1.0 µl/l ozone for 1-5 hr, the mitochondria extracted from the tissue prior to symptom development exhibited reduced oxygen uptake and reduced oxidative phosphorylation (7). In an experiment of similar design when ozone was bubbled through a solution of isolated mitochondria, both respiration and oxidative phosphorylation were reduced (7). As will be elaborated subsequently, it appears that the effects of ozone on mitochondria are secondary.

Studies investigating the photosynthetic function of plants have demonstrated an immediate decrease in photosynthetic rates followed by a return to normal when symptoms appear (8, 9).

This data has been accrued from oxygen evolution and carbon
dioxide fixation measurements conducted on intact leaf tissue
harvested from plants exposed to ozone in vivo. Unfortunately,
isolation of chloroplasts from the leaf tissue was not conducted
to determine photosynthetic or photophosphorylation rates.
Therefore, we cannot compare the photosynthetic data secured from
leaf tissue determinations and the respiration data obtained
from studies conducted on the mitochondria. Such comparisons
in situ are not possible because a change in gaseous exchange in
leaf tissue could be a direct result of impaired organelle
function or an indirect effect of altered gaseous ingress due to
another cellular malfunction. Several in vitro studies have been
conducted with isolated chloroplasts in which the organelles were
exposed to ozone; photophosphorylation rates and photosynthetic
parameters were measured. It was demonstrated that when isolated
spinach chloroplasts were exposed to a total of 900 nanomoles or
less of ozone, electron transport and cyclic photophosphorylation
were reduced (10). In another experiment when 40 µl/l ozone
were bubbled through a pea chloroplast suspension, ATP formation
was reduced (11).

Adenosine triphosphate is utilized in portions of the cell
other than the mitochondria and chloroplasts; therefore, the
utilization as well as the production of ATP is of importance to
total adenylate status. As a result, it became important to
consider total ATP content of plants. When detached pinto bean
leaves were exposed to 1.0 µl/l ozone for 30 min total ATP
content of the leaf decreased (12). Since ozone altered leaf
ATP content it could also alter the leaf's adenylate status; we
wished to determine if a correlation existed between alteration
in adenylates and the change previously reported in photosynthesis and respiration. Since ATP is readily broken down by
adenosine triphosphatases, a reliable method of extraction and
quantitative method of ATP analysis was designed for the study
(8).

Adenosine triphosphate was measured by the luciferin-
luciferase assay and total adenylates were measured by converting
adenosine diphosphate (ADP) and adenosine monophosphate (AMP) to
ATP and measuring similarly. Photosynthesis and respiration were
measured manometrically (8). Intact pinto bean plants were exposed to 0.30 µl/l ozone for three hr and analyzed immediately
after ozonation before visual symptoms developed; we observed a
significant increase in ATP and total adenylate content, an increase in respiration rates and a decrease in photosynthesis
rates (Table I). Due to the seeming disparity between our
results and those reported previously (6, 7, 10, 11, 12), we repeated our experiments measuring these parameters at 0, 6, 21,
and 72 hr after ozonation to more thoroughly characterize the
responses with regard to symptom development. We again observed
the significant increase in ATP in total adenylate content and
in respiration rate, and a decrease in photosynthesis immediately

Table I.   ATP and total adenylate content and photosynthesis and respiration rates of pinto bean leaves harvested immediately after a 3-hr exposure to 0.30 µl/l ozone[1]

| | ATP | | Total Adenylates | | Photosynthesis | | Respiration | |
|---|---|---|---|---|---|---|---|---|
| | Control | Ozone | Control | Ozone | Control | Ozone | Control | Ozone |
| | nmole/g fresh wt | | | | µl $O_2$/hr-mg dry wt | | | |
| | $112.4^2$ | $139.7^{xx}$ | 177.7 | $205.7^{x}$ | 4.25 | 3.60 NS | 3.43 | 3.38 NS |
| | 117.4 | $142.7^{xx}$ | 147.9 | $194.4^{xx}$ | 5.83 | 5.25 NS | 4.33 | 4.48 NS |
| | 111.5 | $130.0^{xx}$ | 149.4 | $171.7^{x}$ | 17.58 | $12.03^{x3}$ | 4.00 | 4.93 NS |
| x LSD 0.05 | | | 21.37 | | 0.67 | | 0.72 | |
| xx LSD 0.01 | 8.28 | | 29.14 | | | | | |

[1] Data from Pell (8).

[2] Each number is the mean of four replicates.

[3] Photosynthesis results of this experiment were significant at the 5% level based on t test.

after ozonation (Table II).

These changes persisted for 6 hr. Twenty-one hr after ozone exposure, the ATP level and photosynthetic rate had returned to normal. Seventy two hr after ozonation, respiration rates were still increased but photosynthesis rates and total adenylate levels had returned to normal. The ATP content of the ozonated bean plants showed a secondary increase over control plants after 72 hr.

There are several observations which can be drawn from these studies. Ozone alters respiration, photosynthesis and phosphorylation rates as well as the general adenylate status of plants. In each research paper discussed above, there were reports of changes in various bioenergetic systems. The direction and magnitude of these changes were not always the same. There are many possible explanations for the spectrum of results reported, including environmental conditions before, during and after ozonation and the use of different experimental methods to test any of the cell functions. However, there are two basic differences in experimental design which may be responsible for the variable results: (a) use of different ozone dosage regimes and (b) use of in vivo or in vitro plant systems.

The dose of ozone, i.e. the gas concentration exposure time, will determine the severity and the rate of symptom development. The rate of development is very important when physiological and/or biochemical analyses are conducted on plant tissue. If different doses are used (as was the case above) the rates of symptom development will be different. The times of tissue harvest for functional analyses will probably not be comparable and, therefore, the results may vary for tissue analyzed at different times during ozone symptom development. Since all the ozone concentrations used in the experiments discussed above lead to cell death, it is likely that the functional changes reported do occur. It is the sequence in which these changes occur that is significant, and this is dependent upon when measurements were made with regard to symptom development. When plant tissue was exposed to high ozone concentrations, rapid induction of cellular and organellar destruction could have occurred. Any analysis conducted subsequent to the ozonation would probably reflect an advanced stage of cell damage. As a result, any early event in damage development would be overlooked. For example, an increase in adenylate concentration could occur early in the development of symptoms followed by a rupturing of the integrity of the coupled membrane system. This uncoupling could lead to increased respiration and decreased oxidative phosphorylation. During rapid symptom development an early physiological change could pass before there was an opportunity for analysis.

Both in vivo and in vitro systems have been utilized to try to elucidate the effects of ozone on vegetation, but one must be cautious about assuming that they are equal. In vitro studies reveal the potential of ozone to injure organelles, enzyme systems

Table II.  Mean values for ATP and total adenylate content and photosynthesis and respiration rates of pinto bean leaves harvested over a 72-hr period following a 3-hr ozonation[1]

| Time hr | ATP | | Total Adenylates | | Photosynthesis | | Respiration | |
|---|---|---|---|---|---|---|---|---|
| | Control | Ozone | Control | Ozone | Control | Ozone | Control | Ozone |
| | nmole/g fresh wt | | | | µl $O_2$/hr·mg dry wt | | | |
| 0 | 82.81[2] | 123.06xx | 113.58 | 159.98xx | 5.05 | 3.95xx | 3.63 | 4.23xx |
| 6 | 83.23 | 126.00xx | 124.55 | 181.54xx | 3.88 | 3.34x | 2.91 | 3.75xx |
| 21 | 91.28 | 99.56 NS | 119.06 | 155.88xx | 3.23 | 3.31 NS | 3.00 | 4.50xx |
| 72 | 65.56 | 81.75x | 120.33 | 112.33 NS | 3.21 | 3.30 NS | 3.58 | 4.43xx |
| x LSD 0.05 | 14.68 | | | | 0.52 | | | |
| xx LSD 0.01 | 19.64 | | 25.90 | | 0.69 | | 0.49 | |

[1] Data from Pell (8).

[2] Each number is the mean of eight replicates.

etc., but the data cannot be extrapolated to the effects of ozone on these systems in vivo because the environment of a subcellular structure and/or system in vivo is quite different from that in an artifical, in vitro, medium. This environment may be crucial to the system's response to the gas. For example, organelles are normally found bathed in cytoplasm and are influenced by the metabolic pools existing therein. If an organelle were damaged by ozone and required precursors from the metabolic pool for repair, this response would not be measured in an in vitro system; in fact, a more severe or different response than occurs in vivo may be measured. One such example deals with the decrease in oxidative- and photo- phosphorylation rates observed when mitochondria and chloroplasts were treated and analyzed in vitro. These decreases could have been due to impairment of structures necessary for phosphorylation or to leakage of necessary precursors from the organelle into the buffered solution. The latter is a distinct possibility since there is evidence of damaged membrane function (13). In fact, it has been reported that when isolated mitochondria are exposed to ozone there is increased leakage of nucleotides directly proportional to length of ozone exposure (14). If metabolites and precursors are leaking into the cytoplasm, consideration should be given to the impact which these metabolites might have on cellular functions. For example, if adenylates leak out of mitochondrial and chloroplast membranes into the cytoplasm of intact cells, a demand for increased synthesis of adenylates could occur within these organelles. This synthesis would then result in an increase in net adenylate content of the entire tissue. The enzymes necessary for this synthesis are believed to be cytoplasmic in origin; hence, if the cytoplasm is not present--as is the case in an in vitro experiment--the increase could not occur.

From the discussion above we conclude that our observations are not necessarily inconsistent with those of other reports. Both the time of analysis and experimental design may affect the results. An explanation for the increase in adenylates under the conditions of our experiment is still needed. Since both ATP alone and total adenylate concentrations have increased, it does not appear that a shift in phosphorylation can account for the increases. The decrease in photosynthesis and increase in adenylates occur during the same time period and both factors return to normal after 21 hr. From previous research we know that the photosynthetic levels of ozonated pinto bean foliage decrease immediately after ozone exposure even when symptoms do not develop (8). This does not hold true for the adenylate or respiration responses. Therefore, it appears that the ozone-initiated increase in adenylates is not correlated directly to the photosynthetic response. The increase in respiration persists when adenylate content and photosynthetic rates have returned to normal. Impaired mitochondrial function appears to be a secondary response more closely related to symptom development.

If photosynthetic and respiratory changes cannot account for the increases in adenylate concentration, which system is responsible? It has been reported that ADP and ATP concentrations of Ehrlich ascites tumor cells increase in the presence of adenine (15). Whether this would hold true for plant cells is not known, but it seems plausible that equilibrium shifts would initiate similar responses. An increase in adenine concentrations could occur if there was any breakdown of nucleic acids. There is one report that the number of ribosomes in the chloroplast does decrease in response to ozone (16). An increase in synthesis of purines is also possible but there is no evidence to either support or refute this hypothesis.

Increases in adenylate content as have been discussed here are not unique responses of plants to ozone. Other stresses, including the air pollutant hydrogen fluoride and tobacco mosaic virus, have been reported to induce similar increases in ATP and/or adenylates (17, 18). Although ATP is considered an energy source and, therefore, a valuable biochemical cell constituent, its production at the expense of other cellular components would have deleterious effects. Furthermore, it is not the total concentrations of adenylates per se, but the distribution which is important. Some researchers have suggested that the movement of adenylates from the production site to synthesis sites controls cellular metabolism (19, 20). If the distribution shifts, metabolism will be altered. Atkinson (2) has stated that the energy charge may be responsible for controlling of gene function. Since there are at least three sites of functional gene activity (nucleic acids) viz. chloroplast, mitochondria and nucleus, the energy charge would be crucial in all these areas. It would be important to determine whether the increase in total adenylates reflects any change in energy charge in any or all sites of gene function. This would be accomplished by measuring absolute concentrations as well as ratios.

The impact of altered energy charge is open to speculation since we do not fully understand the role of energy charge in the healthy cell. Many researchers have suggested that ozone induces premature senescence (21, 22). Normal senescence is a function of plant growth related directly to gene function. Perhaps ozone-induced senescence could be attributed to the effects of altered adenylate status on gene function.

## Literature Cited

1.  Dugger, W. M. and Ting, I. P.  Ann. Rev. Plant Physiol.
    (1970) 21:215.
2.  Atkinson, D. E. In "Horizons of Bioenergetics."  A. San
    Pietro, Gest, H. Eds., pp. 83-97.
3.  Atkinson, D. E.  Ann. Rev. Biochem. (1966) 35:85.
4.  Todd, G. W.  Plant Physiol. (1958) 33:416.
5.  Todd, G. W. Physiol. Plantarum (1963) 16:57.
6.  Macdowall, F. D. H.  Can. J. Bot. (1965) 43:419.
7.  Lee, T. T. Plant Physiol. (1967) 42:691.
8.  Pell, E. J. and Brennan, E.  Plant Physiol. (1973) 51:378.
9.  Hill, A. C. and Littlefield, N. Environ. Sci. Technol.
    (1969) 3:52.
10. Coulson, C. and Heath, R. L.  Plant Physiol. (1974) 53:32.
11. Nobel, P. S. and Wang, C. Arch. Biochem. Biophys.
    (1973) 157:388.
12. Tomlinson, H. and Rich, S. Phytopathology (1968) 58:808.
13. Spotts, R., Lukezic, F. and Hamilton, R.  Phytopathology
    (1974) in Press.
14. Lee, T. T. Plant Physiol. (1968) 43:133.
15. Snyder, F. F. and Henderson, J. F.  Can. J. Biochem.
    (1973) 51:943.
16. Chang, C. W.  Phytochemistry (1971) 10:2863.
17. Lords, J. L. and McNulty, I. B.  Utah Acad. Sci. Arts Lett.
    (1965) 42:163.
18. Sunderland, D. W. and Merrett, M. J. Physiol. Plantarum
    (1967) 20:368.
19. Heldt, H. W. FEBS Letters (1969) 5:11.
20. Santarius, K. A. and Heber, U. Biochim. Biophys. Acta
    (1965) 102:39.
21. Heggestad, H. E. Amer. Potato J. (1973) 50:315.
22. Rich, S. Ann. Rev. Phytopath. (1964) 2:253.

# 10

# Acute Inhibition of Apparent Photosynthesis by Phytotoxic Air Pollutants

JESSE H. BENNETT

U. S. Department of Agriculture, Beltsville, Md. 20705

A. CLYDE HILL

University of Utah, Salt Lake City, Utah 84112

## Abstract

Sublethal plant exposures to a number of phytotoxic air pollutants can cause the reversible suppression of one of life's most basic processes-- photosynthesis. The possibility of plant growth suppression by atmospheric pollution is a concern of many people. We need to know if subnecrotic pollutant exposures that may occur in ambient air can repress photosynthesis rates sufficiently to cause significant retardation of plant growth. Some insight into the capability for short-term exposures to HF, $Cl_2$, $O_3$, $SO_2$, $NO_2$ and NO-- applied singly and as dual pollutant mixtures-- to suppress apparent photosynthesis rates of several important crop species is presented here.

## Introduction

Many agriculturists express the conviction that crop growth and productivity can be suppressed by phytotoxic air pollutants present in the atmosphere in dosages that may be too low to cause obvious symptoms of foliar injury. The mechanisms of growth suppression are not well understood; however, adverse effects on a number of plant functional and structural processes could lead to growth reduction. An important in vivo process that can be inhibited by plant exposure to certain air-polluting phyto-toxicants is the plant's net photosynthesis rate ([1-9]). Several sensitive and rapid methods for detecting effects on photo-synthesis are available. Three of these involve (a) determining the rates that the physiological gases $CO_2$ or $O_2$ are exchanged with the plants, (b) measuring leaf ATP and (c) investigating changes in the fluorescence of photosynthesizing tissues. Carbon Dioxide exchange measurements have been utilized as a screening technique to determine which important air pollutants can inhibit photosynthetic rates in plants exposed to low pollutant concen-trations and exposure dosages ([1-3]).

Phytotoxic atmospheric pollutants have been rated by plant

115

scientists as to their relative importance in affecting plant
life in the United States (10,11). Those recognized as being of
greatest significance are: ozone and other oxidants of the
photochemical oxidant complex (i.e. peroxyacetyl nitrates and
nitrogen oxides), sulfur dioxide, agricultural chemicals,
fluorides and ethylene. Other atmospheric pollutants of lesser
significance include chlorine and hydrogen chloride, acid
aerosols, and heavy metals and radioactive substances. Problems
caused by air pollution vary in both severity and nature in
different parts of the country. Many factors contribute to this
situation. These include the particular pollutant sources
present; the accumulation, interaction and fate of the pollutants
in the environment; the kinds, distribution and use of the plants;
and the relative sensitivities of vegetation growing in their
respective habitats.

Ozone ($O_3$), the most ubiquitous air pollutant in the U.S.,
is rated as the single most important phytotoxic pollutant affecting vegetation (10-12). The highest concentrations of $O_3$ (and
photochemical oxidants) in the country are found in Southern
California. Ozone does occur in elevated concentrations over
extensive areas of the heavily urbanized Eastern United States,
however, and greater total losses due to vegetation injury may
be sustained there. It should be noted that plants growing in
humid areas characteristic of the Eastern U.S. tend to have increased sensitivities to phytotoxic air pollutants in comparison
to those grown in drier habitats.

Sulfur dioxide ($SO_2$) is regarded as the most important phytotoxic air pollutant emitted from industrial (point) sources.
Sulfur dioxide is commonly formed during the combustion of sulfur-
containing fuels (especially coal and petroleum) and from smelt-
ing operations using sulfide ores. Nitrogen oxides also arise as
a result of these high temperature processes and can be emitted
simultaneously. Some studies indicate that the phytotoxicity of
air containing $SO_2$ and $NO_2$ is enhanced when both pollutants are
present together and concern has arisen recently about the
quality of air possessing elevated concentrations of these pollutants (9, 13). Likewise, ambient air containing $O_3$ into which
$SO_2$ is emitted has become suspect since $SO_2$ and $O_3$ have been
reported to interact synergistically to injure plants (12).

Information presently available will allow only a first
approximation to be made into the significance of air pollution
on plant growth. To complicate an evaluation, very little is
known about the impact of various pollutant combinations that
might interact to determine the phytotoxicity of the mixtures.
Current knowledge about the inhibitory effects of major air
pollutants on apparent photosynthetic rates in plants is described
below. The discussion includes observed plant responses to
simple dual combinations of the pollutants.

## Single Pollutant Exposures

At least six major phytotoxic air pollutants have been shown
to reversibly inhibit apparent photosynthetic rates in plants
(1-3). Studies indicate that these phytotoxicants ranked in the
following order according to the relative amount of inhibition
effected after several hours of exposure to equal pollutant
concentrations: $HF>Cl_2 \approx O_3>SO_2>NO_2>NO$. A summary of the experimental results which compares measured depressions in $CO_2$ uptake
rates of barley and oat canopies after 2-hr pollutant exposures
in environmental chambers appears in Figure 1. Typical inhibition and recovery rate curves for exposures that reduced $CO_2$
absorption rates by 20 percent at the end of the 2-hr fumigations
are also shown. Similar data have been obtained for alfalfa,
another important crop species which was cultured and exposed
under identical conditions (2,3). In contrast, equivalent
experiments on a plant species (Muhlenbergia asperifolia) that
is very resistant to $SO_2$ showed this grass to require about five
times greater $SO_2$ exposure dosages in order to induce comparable
rate reductions (3). Responses of many plants subjected to
similar environmental treatment conditions would probably fall
between this 1:5 sensitivity range. Plants grown and exposed
under more adverse (i.e. xeric, etc.) cultural circumstances
tend to respond less markedly to given pollutant dosages than
plants subjected to environments which foster rapid growth. Such
plants may appear to be relatively tolerant to pollution. Plant
age and stage of development also greatly influence plant tissue
sensitivity to pollutant-induced injury. The plant canopies
referred to above were composed of actively growing 3-5 week-old
plants or 3-5 weeks of regrowth after cutting 1-3 times in the
case of alfalfa.

Since many factors that regulate the gas and energy balance
of plants influence plant photosynthesis and other responses to
air pollutants, experimental conditions were evaluated to provide
microenvironmental parameters that could permit reasonable plant
photosynthesis and pollutant absorption rates which might occur
in the field (1,14). The environmental chamber systems developed
allowed canopy $CO_2$ and $O_3$ uptake rates that averaged more than
85 percent of the rates determined for plants exposed to full
sunlight. Sulfur dioxide uptake rates by alfalfa in the chambers
were slightly higher (by about 20 percent) than $^{35}SO_2$ uptake by
an equivalent alfalfa canopy exposed under ambient conditions in
the field. Table I gives chamber conditions. The microclimate
surrounding each standardized experimental plant canopy was
thought of as a six-dimensional space in which the most important
interacting microenvironmental variables--radiation, wind, air
temperature, relative humidity, and $CO_2$ and pollutant concentra-
tions--were controlled at selected values. Changes in canopy
$CO_2$ uptake rates from pre-established steady-state levels were
determined as time-dependent functions.

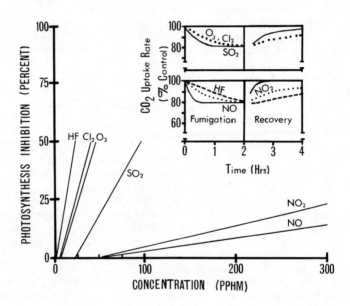

*Figure 1.    Inhibition of apparent photosynthetic rates of barley
and oat canopies by 2-hr air pollutant fumigations.*

Table I.  Environmental chamber conditions

| Variable | Chamber Values* |
|---|---|
| 1. Light intensity | 40 - 50 Klux |
|     Source: | |
|     a. Lamps (fluorescent, tungsten and quartz iodide lamps) | total power:7665 watts |
| 2. Wind velocity | 1.2 - 1.6 m/sec. |
| 3. Air temperature | 24 ± 2°C |
| 4. Relative humidity | 45 ± 5% |
| 5. $CO_2$ concentration | 300-320 pphm |
| 6. Pollutant concentration | varied with experiment |

* Values given describe conditions in the chamber air space above the canopies except for light intensity (40-50 Klux) which is for the average canopy height (top) level.

Given these restrictions, oat, barley and alfalfa canopies required treatments with more than 1 pphm HF, about 5 pphm $O_3$ or $Cl_2$, 20 pphm $SO_2$, and 40-60 pphm $NO_2$ or NO before apparent photosynthesis rates were measurably depressed by the end of 2 hr of exposure.  Above these apparent threshold values the 2-hr depressions induced were linearly related to pollutant concentrations applied up to those that caused visible foliar injury to the tissues.  Foliar necrosis occurred to some plant tissues within the canopies exposed to approximately 15 pphm HF, 20 pphm $O_3$ or $Cl_2$, 80 pphm $SO_2$ and more than about 500 pphm $NO_2$. Nitric oxide did not visibly injure plants in any experiment conducted.  Some tissue destruction occurred in plants given HF, $Cl_2$, $O_3$ and $SO_2$ which reduced overall canopy $CO_2$ uptake rates by 25-50 percent.  (Although, apparent photosynthesis rates may have been more severely depressed in the individual leaves that were visibly injured.)  Essentially complete inhibition of canopy $CO_2$ uptake rates could be temporarily produced by nitrogen oxide, however, without producing observable tissue destruction ([2]).

Apparent photosynthetic rates in plants subjected to $SO_2$ or NO exposures with constant pollutant concentrations, as illustrated in Figure 1, characteristically dropped rapidly upon initiation of treatment to new depressed equilibrium levels which could be maintained for several hours.  Hydrogen fluoride, conversely, caused $CO_2$ uptake rates to decline more gradually during fumigation.  Chlorine, $O_3$ and $NO_2$ exposures induced inhibition rate responses which were intermediate between these extremes.

   After fumigation, plants previously exposed to NO or $SO_2$ re-covered much more quickly than plants given HF exposures. Plants treated with $Cl_2$, $O_3$ or $NO_2$ recovered at intermediate rates. Hydrogen fluoride-suppressed plants exhibited a short lag period after exposure which was followed by a relatively slow recovery rate. The gradual reduction in $CO_2$ absorption rates and slow recovery by HF treated plants may indicate fluoride accumulation in the leaf tissue with increasing length of exposure followed by relatively slow removal or leaf detoxification and repair after termination of treatment. This gas is highly soluble in water and $F^-$ binds with many cellular constituents. By contrast, NO is only slightly soluble in aqueous media and the cellular reaction products are thought to be transient (2). [It has been suggested that iron-containing redox agents which function in photosynthesis energy-transfer processes (e.g., ferredoxin) might be loosely complexed by NO (interfering with electron transport), with the iron-NO complexes dependent upon NO concentrations in the cellular solutions.] Nitric oxide tends to obey Henry's Law -- that the concentration of gas dissolved in an aqueous medium may be expected to be proportional to the partial pressure of the external gas if the solution is ideal or nearly so-- and NO in the cellular medium may be readily influenced by changes in its atmospheric partial pressure. Cellular fluoride would be comparatively uncoupled from its HF partial pressure. The plant responses observed illustrate predictable consequences of pollutant solubility and reaction properties on cellular absorbed dosage and effects.

   The ready modulation of $CO_2$ uptake rates by $SO_2$ changes in the atmosphere suggest that phytotoxic cellular sulfite concentrations may be closely coupled to the gaseous partial pressure of $SO_2$. The moderately slow inhibition and recovery rates induced by $NO_2$ may be due in part to its tendency (a) to decompose in solution forming nonvolatile nitrite (and nitrate) which might accumulate in cells during exposure with concentrations high enough to depress $CO_2$ uptake rates, and (b) to be depleted relatively slowly through metabolic or translocation processes after fumigation. Excess nitrite may inhibit $CO_2$ fixation by interfering with ferredoxin mediated reductive processes (2).

   Though chemically different, dissolved sulfur and nitrogen oxides show similar properties in biological systems. Both can serve as electron acceptors in metabolic reactions and involve a maximum of 8-electron changes each. The reduced forms can be incorporated into organic compounds or be reoxidized in the cells with part of the energy being conserved in phosphate bond formation. There is also evidence that sulfite and nitrite may be reduced by the same or closely associated enzyme systems in some organisms studied (15-19).

   Plant responses to $Cl_2$ and $O_3$ exposures were similar with regard to the inhibition and recovery rate curves produced and the overall phytotoxicities resulting from equal exposure dosages. Although $O_3$ is a more powerful oxidant than $Cl_2$, chlorine is more soluble in aqueous media and may be absorbed faster by leaves ([14]). These potent oxidants can react with cellular membranes and compounds, leading to alterations in the functional and structural integrity of the cells. These two pollutants also cause leaf stomata to close under certain conditions and can affect $CO_2$ exchange rates by impeding the diffusion of $CO_2$ from surrounding air into the leaves ([1-3]). Transpiration rate measurements indicated that $O_3$ and $Cl_2$ increased leaf resistances to gas transfer in these experiments in proportion to the amount of suppression induced. The reversible inhibition observed may have resulted largely from the increased resistance to gas diffusion.

This was not found to be the case for the nitrogen oxides. Transpiration rates were not depressed significantly by exposures to NO or $NO_2$ that caused marked depressions in the $CO_2$ uptake rates. The nitrogen oxides appeared to inhibit $CO_2$ uptake rates by affecting the biochemical fixation of $CO_2$.

Although stomata in plants treated with HF and $SO_2$ showed some tendency to close as a result of exposures which depressed apparent photosynthetic rates, these phytotoxicants inhibited $CO_2$ uptake rates more by affecting biochemical processes within the leaves than by impeding gas transfer by inducing stomatal closure.

Another important phytotoxic atmospheric pollutant that has been studied with respect to its inhibitory effects on plant photosynthesis is peroxyacetyl nitrate (PAN). This phytotoxicant applied for 30 min at 1 ppm depressed the incorporation of $^{14}CO_2$ into intact pinto bean leaves, but only after visible tissue injury started to develop ([20]). From companion studies on isolated chloroplasts, it was concluded that PAN-induced inhibition was probably associated with the carboxylating reaction or the chloroplast light-energy conversion system leading to assimilative power. The inhibition appeared to result in a quantitative reduction (but not a qualitative change) in the early products of photosynthesis.

The possible effects of carbon monoxide on alfalfa canopies have also been tested but CO did not measurably depress $CO_2$ uptake rates when present in concentrations ranging up to 80 ppm in the fumigant ([3]).

## Exposures to Pollutant Combinations

Situations where more than one phytotoxicant is present in the atmosphere commonly occur and the possibility of synergistic (greater than additive) or antagonistic (less than additive) interactions resulting from the pollutant combinations should be recognized. The potentiation of injury symptoms by plant exposures to mixtures of $SO_2+O_3$ and $SO_2+NO_2$ has been referred to previously. One recent study (9) showed evidence that $SO_2+NO_2$ mixtures can depress the apparent photosynthetic rates of alfalfa more than would be expected from summing the observed depressions experimentally induced by treatments with each pollutant alone. Results of another study (2) indicated $NO+NO_2$ combinations were additive in their combined suppression of $CO_2$ uptake rates by oats and alfalfa. Inhibitory effects induced by simple dual combinations of several important phytotoxic pollutants which can suppress $CO_2$ uptake by plants are summarized in this section (Table II).

Table II includes supporting data for greater-than-additive inhibition of alfalfa apparent photosynthetic rates induced by $SO_2+NO_2$ mixtures. The enhanced effects were most marked at the lower concentrations applied, becoming less pronounced as pollutant levels were raised. At 50 pphm of each gas no synergism was evident. At this $SO_2$ exposure concentration, sulfur dioxide appeared to regulate the observed plant responses. Significant amounts of inhibition resulted from the lowest bipollutant concentrations used (15 pphm of each gas); these concentrations were well below those required for the individual pollutants to measurably suppress apparent photosynthesis rates. At these exposure levels where no tissue necrosis occurred, the plants recovered completely within 2 hr after fumigation. The manner by which this inhibiting interaction occurred is not well understood. This pollutant combination is also known to act in a synergistic fashion to cause visible injury to plants, and further study of this mixture may be warranted.

All other pollutant combinations tested with the possible exception of $SO_2$ + HF caused plant responses which were additive in suppressing apparent photosynthesis rates. No evidence of antagonism was observed for any of the dual pollutant mixtures examined. Table II, in addition to summarizing the inhibitory effects caused by $SO_2+NO_2$ mixtures, presents results of alfalfa exposures to $O_3$, $SO_2$, HF, $NO_2$ and to the following combinations: $SO_2+O_3$, $SO_2+HF$, $NO_2+HF$ and $O_3+NO_2$. Pollutant concentrations used in these investigations were near the lowest concentrations which measurably suppressed plant $CO_2$ uptake rates but they represent moderate to high concentrations with regard to those that are generally found in polluted ambient air.

In real atmospheres a wide array of pollutant combinations may occur. Plant responses described here represent only experimental combinations of major pollutants shown to inhibit $CO_2$ absorption rates. Effects of other important phytotoxic atmospheric pollutants such as ethylene should also be examined along with more complex mixtures. Information regarding the responses of a wider range of plants subjected to varied environmental conditions would further aid in clarifying the problem.

Table II.  Inhibition of apparent photosynthetic rates of alfalfa
by 1-hr treatments with $O_3$, $SO_2$, HF, $NO_2$, and dual
combinations of the pollutants

| Pollutant(s) | Conc. (pphm) | n | $\overline{\Delta P}$(% Control)[1] | $\overline{\Sigma \Delta P}$(% Control)[2] |
|---|---|---|---|---|
| $O_3$ | 10 | 5 | 4 ± 3 | -- |
| | 20 | 5 | 10 ± 4 | -- |
| $SO_2$ | 15 | 5 | 0 | -- |
| | 25 | 15 | 2 ± 1 | -- |
| | 30 | 5 | 6 ± 3 | -- |
| | 50 | 10 | 21 ± 3 | -- |
| $NO_2$ | 25 | 13 | 0 | -- |
| | 40 | 5 | 2 ± 2 | -- |
| | 50 | 5 | 3 ± 3 | -- |
| HF | 3 | 5 | 3 ± 2 | -- |
| $SO_2$+$NO_2$ | (15 + 15) | 5 | 7 ± 2** | 0 |
| | (25 + 25) | 13 | 9 ± 2** | 2 ± 1 |
| | (50 + 40) | 7 | 20 ± 4 | 23 ± 3 |
| $SO_2$+$O_3$ | (30 + 10) | 5 | 11 ± 3 | 10 ± 4 |
| | (30 + 20) | 5 | 19 ± 4 | 16 ± 5 |
| $SO_2$+HF | (25 + 3) | 5 | 9 ± 3* | 5 ± 2 |
| $NO_2$+HF | (50 + 3) | 5 | 7 ± 3 | 6 ± 4 |
| $NO_2$+$O_3$ | (50 + 10) | 3 | 9 ± 4 | 7 ± 4 |

[1] Mean depression in $CO_2$ uptake rates induced after 1-hr
exposures, and 95% confidence interval.  (Superscripts ** and
* denote those significant means at the P.01 and P.05 levels,
respectively, when compared with the summed depressions determined
for the separately applied pollutants at corresponding concentrations.)

[2] Summed duo means ($\overline{\Delta P_i} + \overline{\Delta P_j}$) and 95% confidence intervals computed
from pooled variances (21,22).

Potential Impact of Air Pollution on Photosynthesis

     Although information concerning the inhibition of photo-
synthesis by air pollution is limited, we may gain perspective
into the potential problem through appraising available data on
the extent that $CO_2$ uptake by oats, barley, and alfalfa canopies
can be suppressed by short-term (a few hours) exposure to the
major air pollutants and simple combinations investigated.

     Of the phytotoxic air pollutants and mixtures tested, $O_3$ or
combinations of $SO_2+NO_2$ are most likely to occur in ambient
atmospheres in sufficiently high concentrations to acutely
depress apparent photosynthesis.  Ambient HF concentrations of
the magnitudes which inhibited $CO_2$ uptake rates in an acute,
reversible manner would be rare.  Studies into longer-term exposures (several days or weeks) to HF concentrations in the low
ppb range have suggested that reduced photosynthesis under these
conditions correlated with the amount of necrosis that developed
(5,6).

     In concentrations approximating present air quality stan-
dards (Table III), $O_3$ or $SO_2$ in combination with $NO_2$ could
measurably suppress $CO_2$ uptake rates of sensitive plants if exposed under favorable growing conditions.  In the controlled
environmental chamber studies, 1-hr exposures to 10 pphm $O_3$
(which is slightly above the primary and secondary standards --
i.e., 8 pphm for 1 hr) for example, depressed alfalfa $CO_2$
absorption rates by approximately four percent.  Exposures to
15 pphm hr $SO_2$ in combination with an equal amount of $NO_2$ reduced
uptake rates by 7 percent.  Alfalfa, barley or oat canopies exposed to these pollutants singly required higher concentrations
(i.e., 1- to 2-hr treatments with more than 20 pphm $SO_2$ or 40
pphm $NO_2$) to measurably reduce canopy uptake rates.


Table III.  Summary of National Ambient Air Quality Standards
                    for the United States

| Pollutant | | Standards (pphm) | |
|---|---|---|---|
| | | Primary | Secondary |
| Sulfur oxides ($SO_x$) | Yearly mean | 3 | |
| | Maximum 24-hour average | 14 | |
| | Maximum 3-hour average | | 50 |
| Oxidants ($O_x$) | Maximum 1-hour average | 8 | 8 |
| Nitrogen oxides ($NO_2$) | Yearly mean | 5 | 5 |
| Hydrocarbons | Maximum 3-hour average | 24 | 24 |
| Carbon monoxide (CO) | Maximum 8-hour average | 9(ppm) | 9(ppm) |
| | Maximum 1-hour average | 35(ppm) | 35(ppm) |

Reports in the literature (12,13) indicate that certain highly sensitive plants may be injured by several hours exposure each day over several weeks' time to low $SO_2+O_3$ concentrations (ca. 5 pphm of each pollutant--a concentration of $O_3$ which is close to natural background levels) or by short-term exposures to concentrations of $SO_2+NO_2$ (5-25 pphm of each pollutant) that have been shown to reversibly inhibit apparent photosynthetic rates of alfalfa.  It might be presumed that the phytotoxic gas mixtures which caused plant injury could have markedly depressed plant $CO_2$ uptake during the exposures.  Sulfur dioxide + nitrogen dioxide exposure dosages that were higher than these (i.e., 3-hr exposures to more than 25-50 pphm $SO_2+NO_2$) were required to cause injury to radish, alfalfa, and several other plants tested in another study (23) conducted under experimental conditions similar to those given in Table I.  The amount of alfalfa $CO_2$ uptake inhibition caused by $SO_2+NO_2$ exposures to the 5-25 pphm concentration range in the investigations presented here was relatively small (<10%).  The experiments on alfalfa canopies using $SO_2+O_3$ mixtures in the 10-20 pphm $O_3$ and 30 pphm $SO_2$ concentration range indicated that apparent photosynthetic rates could be depressed as much as 20-25% by 1-hr fumigations at the higher levels and that the effects were additive (Table II).

Effects of chronic or repeated exposures to subnecrotic levels of phytotoxic pollutants and their pollutant mixtures need to be considered.  Recurring sublethal exposures of short duration might modify plant responses to succeeding episodes, but experimental data presently available suggest that some plants can recover repeatedly from treatment with sulfur and nitrogen oxides provided sufficient time is allowed between fumigations for full recuperation (3,19).  It is possible that prolonged pollutant exposures could chronically depress plant photosynthesis.  Some cellular destruction generally results to foliar tissues before canopy $CO_2$ absorption rates are greatly reduced by $O_3$, $SO_2$, HF, $Cl_2$, and probably their mixtures, however, and visible injury symptoms might be expected to occur if photosynthesis were repressed sufficiently to markedly reduce plant growth.

Plants live in a world comprised of numerous limiting and potentially destructive forces.  A developing biosystem is subjected to the integrated influence of all positive and negative forces that interact simultaneously upon the system.  In the complex natural environment, a practical evaluation of the impact of any one factor should not be made without due respect given to the relative influences of other controlling and limiting factors which can vary plant responses and perturb the system.  It is in this context that the role played by air pollution in affecting our plant resources should be evaluated in the future.

## Literature Cited

1. Hill, A. C. and Littlefield, D.  Environ. Sci. Technol. (1969) 3:52-56.
2. Hill, A. C. and Bennett, J. H.  Atmos. Environ. (1970) 4:341-348.
3. Bennett, J. H. and Hill, A. C.  J. Environ. Quality (1973) 2:526-530.
4. de Koning, H. W. and Jegier, S.  Atmos. Environ. (1968) 2:1-6.
5. Hill, A. C.  J. Air Poll. Cont. Assoc. (1969) 19:331-336.
6. Thomas, M. D.  Agron. J. (1968) 50:545-550.
7. Thompson, C. R., Taylor, O. C., Thomas, M. D. and Ivie, J. O.  Environ. Sci. Technol. (1967) 1:644-650.
8. Todd, G. W. and Propst, B.  Physiol. Plantarum (1963) 16:57-65.
9. White, K. H., Hill, A. C. and Bennett, J. H.  Environ. Sci. Technol. (1974) 8:574-576.
10. Heck, W. W., Taylor, O. C. and Heggestad, H. E.  J. Air Poll. Cont. Assoc. (1973) 23:257-266.
11. Heggestad, H. E.  In W. H., Smith, F. E. and Goldberg, E. D. (eds.) "Man's Impact on Terrestrial and Oceanic Ecosystems." pp. 101-115.  The Massachusetts Institute of Technology Press, Cambridge, Mass. (1971).
12. Heggestad, H. E., Anderson, C. E. and Feder, W. A.  Paper 74-224, presented at the 64th Ann. Meeting of the Air Poll. Cont. Assoc., Denver, Colo., June 1974.
13. Tingey, D. T., Reinert, R. A., Dunning, J. A. and Heck, W. W.  Phytopathol. (1971) 61:1506-1511.
14. Hill, A. C.  J. Air Poll. Cont. Assoc. (1971) 21:341-346.
15. Asada, K. Tamura, G. and Bandurski, R. S.  J. Biol. Chem. (1969) 244:2904-2915.
16. Bandurski, R. S.  In Bonner, J., and Varner, J. E., (eds.). "Plant Biochemistry." pp. 467-490.  Academic Press, New York (1965).
17. Cresswell, C. F., Hageman, R. H., Hewitt, E. J. and Hucklesby, D. P.  Biochem. J. (1965) 94:40-53.
18. Kemp, J. D., Atkinson, D. E., Ehret, A. and Lazzarini, R. A.  J. Biol. Chem. (1963) 238:3466-3471.
19. Thomas, M. D., Hendricks, R. H. and Hill, G. R.  Ind. Eng. Chem. (1950) 42:2231-2235.
20. Dugger, W. J., Jr., Koukol, J., Reed, W. D. and Palmer, R. L.  Plant Physiol. (1963) 38:468-472.
21. Snedecor, G. A., and Cochran, W. G.  "Statistical Methods." pp. 91-134, The Iowa State Univ. Press, Ames, Iowa.  6th Ed. (1972).
22. Longley-Cook, L. H. "Statistical Problems."  pp. 244-274. Barnes and Noble, Inc. N. Y. (1970).

23.  Bennett, J. H. and Hill, A. C.  Paper #0637.  Presented at
     the 2nd International Congress of Plant Pathology,
     Minneapolis, Minn.  September 10, 1973.

# An Early Site of Physiological Damage to Soybean and Cucumber Seedlings following Ozonation

HUGH FRICK and JOE H. CHERRY

Department of Horticulture, Purdue University, West Lafayette, Ind. 47907

Abstract

The cucumber cotyledon and first leaf, and the soybean first leaf and trifoliate leaf are sensitive to ozone concentrations (24 and 50 pphm, respectively, for a 4-hr exposure) within the range expected under adverse atmospheric conditions throughout their rapid phases of organ growth. Chlorophyll content and fresh weight are reduced in these tissues 24 hr after ozonation in a non-linear fashion dependent upon both dose and the stage of organ development. Developmentally repeating organs appear to repeat the response to ozone.

Short-term ozone exposures (45 ± 5 pphm, 15 min) of seedlings were followed after 24 hr by non-linear reductions in chlorophyll/g fresh weight and stimulations in fresh weight/organ. The utilization of $^{14}$C-protein hydrolysate by tissue discs is not only predominantly energy-dependent, but also strictly dependent with respect to inhibition or stimulation upon the time after ozonation. Uptake of labelled amino acids into the soluble pools of tissue discs is sensitive to as little as 15 min of in vivo exposure (45 ± 5 pphm), and incorporation into insoluble protein is sensitive during 15 to 30 min, after which no further reduction is observed for up to 90 min of exposure. The reduction of amino acid influx into the soluble pools is not accountable to a reduction in amino acid tRNA charging, and is probably not due to a reduction in amino acid incorporation. These early events of ozone damage are discussed with respect to elapsed time between onset of ozonation and assay.

## Early Events in Ozone-Induced Plant Injury

The economic significance of air pollutant damage to a wide range of plants of agricultural interest has been increasingly established in recent years (1-6). The necessity to develop operational methods of reducing such damage becomes overt in view of the continual growth of urban sources of primary and secondary

atmospheric pollutants (7-9). Feasible programs of protection
will depend upon clear information about the ways developing
plant systems interact with these pollutants (10-12) and,
ideally, knowledge of the primary site or process protectable.

The effects of the secondary air pollutant, ozone, on
growing plant systems have been extensively documented phenomeno-
logically. There are at least three proposed mechanisms by which
ozone could cause injury, each of which must operate within a
framework of temporal sensitivity determined by the state of
differentiation of the plant (11).

The strong oxidizing potential of free ozone is sufficient
to chemically alter protein amino acids both in vitro (13) and
in vivo (14, 15); the aromatic amino acids are particularly
sensitive in vitro (13). Correlations between ozone sensitivity
and lowered (or perturbed) free amino acid (14, 15), sulfur (16),
nitrogen (17-19) and carbohydrate (14, 20, 21) contents have been
observed. It has also been reported that ozone can indirectly
oxidize proteinaceous sulfhydryl groups and inhibit chloroplast
glycolipid biosynthesis, presumably by first oxidizing chloroplast fatty acids (22-26). These observations have been generalized into the proposal that the ability of ozone to induce
measurable injury is a function of its ability to biologically
titrate reducing substances (27, 28) beyond some threshold value
where damage becomes irreversible (29).

The time sequence of ozone effects on processes occurring
within the organelles, and the loss of structural integrity of
the organelles themselves, have suggested a second possible
mechanism. Both photosynthetic and respiration rates, as well
as ATP content, are modified in response to ozone exposure (28,
30-34), though not consistently to the same extent, timing or
direction. A variety of enzymatic activities appears to be
ultimately sensitive to oxidant damage (15, 35-37). The induction
by ozone of visible pigment bleaching and tissue necrosis has
been correlated with progressive destruction of organellar integrity (2, 38, 39). The time-courses of these responses, however,
imply some earlier event(s) of ozone action (38, 40, 41).

Even though subsequent development of visible injury may be
as much a property of basic plant metabolism (44), it is necessary
that stomata be open during exposure to the gas (42, 43) in order
for leaves otherwise potentially ozone-sensitive to manifest
injury symptoms. Thus, after the primary tissue barrier to gaseous exchange is traversed, the second barrier is the plasma
membrane. Recent work has shown that ozonation can disrupt ion
and water flux at the plasma membrane (45, 46). While membrane
functional disturbance need not be indiscriminate (25, 46), the
cited work supports the suggestion that cellular membranes represent a principal locus of ozone action (10, 39, 47, 48).

We have been utilizing soybean and cucumber seedlings in
attempts to acquire physiological evidence of rapid or early
damage symptoms induced by ozone concentrations in a range to be

expected under adverse atmospheric conditions. The present work deals with characterization of temporal changes in sensitivity of seedlings on the bases of fresh weight and chlorophyll content, and with the rapid effects following ozonation upon the utilization of radiolabelled amino acids.

## Materials and Methods

Growth of Seedlings. Cucumber (Cucumis sativus L., cv. Wisconsin SMR-18) and soybean (Glycine max L. Merr., cv. Wayne) were grown in vermiculite and watered daily in a controlled environment chamber (26 ± 2°C day, 21 ± 2°C night) on a 12 hr photoperiod (eight 30-W cool-white fluorescent tubes plus four 25-W incandescent bulbs; 200+foot-candles). The balanced nutrient solution of Frick and Mohr (49), modified to include 1 g/l Ca(NO₃)₂, but without sucrose, succinic acid and kinetin, was applied on alternate days.
Chlorophyll was measured after Arnon (50) in 80% (v/v) acetone extracts buffered with 5 mM TRIS-HCl at pH 7.0.

Ozonation. Seedlings of appropriate age were placed in Mylar-covered fumigation chambers fitted with charcoal "scrubbers" for continuous filtration of ambient influx and efflux gas. Ozone was generated by passing oxygen over electrically-discharging neon tubes having incomplete circuitry. Its concentration was regulated through gaseous flow and the voltage applied, and monitored with a Mast 724-2 ozone meter equipped with ancillary print-out. Total oxygen influx to the ozone generator approximated 200 ml/min, or less than 0.1% of the chamber volume per min. Plants were exposed to ozone beginning at 3-4 hr into the light cycle. For short-term exposure, the required chamber concentration was obtained before plants were introduced, necessitating a 1-2 min drop of about 5 pphm ozone.*

Amino Acid Utilization; Free-Space Washing; Uptake Linearity. The utilization of $^{14}$C-reconstituted algal protein hydrolysate (Schwartz/Mann) by treated cucumber and soybean seedling tissue was measured using 10 mm discs taken immediately following ozonation. The discs were cut into wedges which were immersed in 5 mM TRIS-HCl (pH 6.9) buffer containing 1.6 µc/ml of the $^{14}$C-labelled protein hydrolysate in the light at 26 ± 1°C. They were washed in running tap water for 5-10 min and homogenized (Ten-Broeck) in cold 80% (v/v) acetone plus 5% (w/v) trichloroacetic acid (TCA/acetone).
The efficiency of tap water washing was monitored using the ion flux compartmental analysis (51) in which the efflux rate of stationary state inorganic ions from plant cells may be analyzed into loss from apparent free-space (surface film, cell walls, intercellular space), from cytoplasm, and from tonoplast.

---

*1 pphm = 0.01 ppm ≈ 0.01 µl/liter.

Discs of control and ozonated soybean seedlings were washed after
4 hr of labelling at 26°C in either tap water or 50 mg/100 ml
casein hydrolysate, and samples of the wash media taken at short
intervals for 90 min.  The $\log_{10}$ of the counts remaining at each
time-point during linear efflux showed that the efflux rate of
apparent free space label did not differ between water and casein
washing, nor did it differ between control and ozonated
(50 ± 5 pphm, 90 min) discs.  Thus, longer than 5 min washing was
sufficient to remove free space label, leaving soluble pool label
and incorporated label.

Similar results were obtained using 12-day-old cucumber leaf
discs washed in tap water;  replicate samples of the discs themselves were taken at short intervals for 60 min.  No change in
incorporated label with time of washing was observed, and no
differences between rates of label loss from soluble pools of
control and ozonated (47 ± 2 pphm, 30 min) plants were observed
even though the ozonation treatment reduced label influx into
the soluble pool by between 35 and 40%.

After centrifugation of the TCA/acetone extract, non-
quenching aliquots of the supernatant were counted by liquid
scintillation, and the pellet digested overnight in 2N NaOH.
Counts present in the TCA/acetone supernatant were taken to represent the uptake of label into the soluble amino acid pool,
while solubilized counts present in the NaOH digest were taken to
represent incorporation into TCA-insoluble protein.

Under these conditions, the time-course of label entry into
both the TCA-soluble pool and the TCA-insoluble protein was
linear for each organ for at least 40 and 50 min, respectively.
It was also linear over these same time periods, although at a
reduced rate, for discs taken from 6-day-old cucumber cotyledons
treated for 2 hr with 50 ± 4 pphm ozone.

## Results and Discussion

Visible Damage Following Ozonation of Cucumber and Soybean
Seedlings; Fresh Weight and Chlorophyll Content Changes.  To
establish for cucumber (cotyledon pair and first leaf) and soybean (leaf pair and first trifoliate leaf) seedlings whether
or not ozone-sensitive stages of growth existed, plants of various
ages and increasing organ maturity were exposed to a fixed dose
and dose-rate of ozone (24 ± 2 pphm for 4 hr for cucumber, and
50 ± 5 pphm for 4 hr for soybean) at hour 3-4 of the light period.
Figures 1, 2, 4 and 5 refer to these experiments using both fresh
weight and chlorophyll content per unit fresh weight as the
criteria of plant injury.

Cucumber Cotyledon and First Leaf.  The cucumber cotyledon
pair (Fig. 1) was sensitive by both criteria to exposure to
moderate doses of ozone during the first three weeks of seedling
growth, including the rapid stage of organ fresh weight increase

through 11 days and the stage of constant fresh weight past 11 days. No differentially sensitive period could be distinguished. The cotyledons rapidly (less than 15 min) developed areas of visible "water logging" on both surfaces which became necrotic during the 24 hr of incubation following treatment. The cotyledons regained turgor in undamaged areas usually within 24 hr.

The effects of ozonation on the first cucumber leaf are shown in Figure 2. Again, the organ was generally sensitive by both criteria (fresh weight and chlorophyll content) throughout the rapid growth phase (9-15 days), and it is shown that the fresh weight of the second and third leaves was also reduced when plants were treated during their rapid growth stages. Seedling ages were subsequently chosen to optimize the ozone response for each organ (see below). In these cases, leaves not yet in their rapid growth stage were also retarded in growth. Thus, it must be stressed that the kind of plant injury data obtained here for the first member of a repeated series of plant organs is reiterated as each new organ develops, and that growth is at all times retarded. Since the cucumber leaves are less succulent than the cotyledons, they did not show the rapid development of tissue "water logging" during ozonation, and tended to develop dense areas of pinpoint bleaching over the leaf surface, as distinct from the necrosis and tissue collapse of the cotyledon.

Dose-Response Curves at Variable Dose-Rate, Fixed Time. The loss of fresh weight during the development of necrosis in cucumber cotyledons (Fig. 1) is apparently accompanied by a proportionately lesser loss of chlorophyll, and this was probably why the reduction in specific-weight chlorophyll content was not more extreme following ozonation at moderate doses. The treatment level of 24 ± 2 pphm ozone for 4 hr was arrived at on the basis of dose-response curves such as those shown in Figure 3. Cucumber seedlings were treated at 9 days from sowing for 4 hr at the given rate, and assayed as before after 24 hr incubation. The exposure to 45 ± 5 pphm ozone resulted in such massive and irreversible collapse of cotyledonary tissue that the subsequent chlorophyll/g fresh weight figure was unrealistically high, apparently through the disproportionate loss of fresh weight. It can also be seen from Figure 3 that a nonlinear dose-response curve was obtained for both first leaf and cotyledon pairs.

At a fixed dose-rate (45 ± 5 pphm), the time of exposure was varied from zero to 90 min, and treated plants incubated a further 24 hr before chlorophyll assay. This test also revealed non-linear responses (Table I) for each organ on the bases of chlorophyll and fresh weight contents using seedlings of sensitive ages.

Soybean First Leaf and First Trifoliate Leaf. The responses of the soybean first leaf pair and first trifoliate leaf to 50 ± 5 pphm ozone are shown in Figures 4 and 5. In the case of

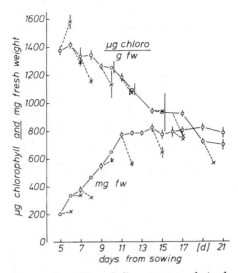

Figure 1. *Chlorophyll content and fresh weight of the cucumber cotyledon pair.* ○ = *untreated plants,* × = *4 hrs at 24 ± 2 pphm ozone and held 24 hrs before assay. Vertical bars indicate standard direction of the mean.*

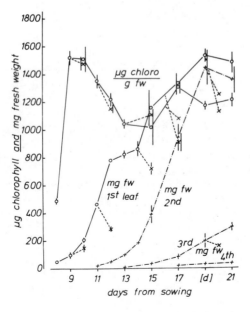

Figure 2. *Chlorophyll content and fresh weight of the cucumber first leaf. Legend as in Figure 1. Fresh weight data for subsequent leaves also shown (- · -).*

Table I.  Response of seedlings to fixed dose-rates of ozone*

| | Time (min) | µg Chl/g fw | mg fresh weight | |
|---|---|---|---|---|
| | | Cotyledon | Cotyledon | Leaf |
| 9d Cucumber | 0 | 1,484 ± 35 | 318 ± 17 | 232 ± 22 |
| | 15 | 1,354 ± 53 | 365 ± 13 | 293 ± 22 |
| | 30 | 1,304 ± 37 | 308 ± 15 | 221 ± 18 |
| | 90 | 1,253 ± 13 | 209 ± 14 | 176 ± 13 |
| | | Leaf | Leaf | Leaf 2 |
| 12d Cucumber | 0 | 1,464 ± 122 | 694 ± 40 | 77 ± 12 |
| | 15 | 1,379 ± 34 | 776 ± 51 | 116 ± 16 |
| | 30 | 1,250 ± 42 | 695 ± 48 | 73 ± 16 |
| | 90 | 1,102 ± 72 | 545 ± 42 | 57 ± 11 |
| | | Leaf | Leaf | Trifoliate |
| 10d Soybean | 0 | 2,132 ± 76 | 341 ± 13 | 153 ± 10 |
| | 15 | 2,060 ± 52 | 310 ± 12 | 132 ± 8 |
| | 30 | 1,951 ± 70 | 318 ± 12 | 138 ± 10 |
| | 90 | 1,895 ± 49 | 268 ± 16 | 114 ± 8 |
| | | Trifoliate | Trifoliate | Trifoliate 2 |
| 14d Soybean | 0 | 1,958 ± 59 | 365 ± 38 | 192 ± 43 |
| | 15 | 2,159 ± 84 | 377 ± 34 | 216 ± 37 |
| | 30 | 2,080 ± 73 | 373 ± 23 | 190 ± 29 |
| | 90 | 1,890 ± 68 | 280 ± 22 | 114 ± 14 |

*Seedlings of the given ages were treated at 45 ± 5 pphm for the lengths of time given and incubated a further 24 hr before fresh weight and chlorophyll assays.

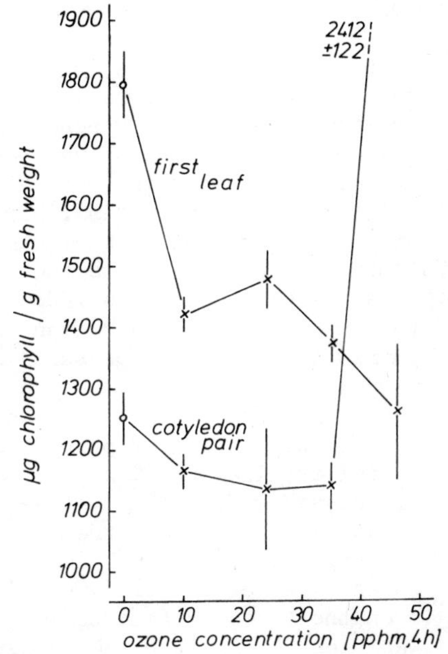

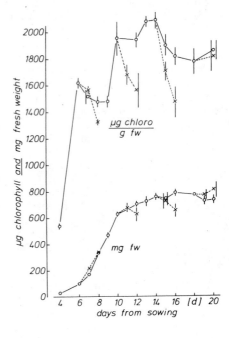

*Figure 3. Ozonation dose-response for 9-day cucumber seedlings. Legend as in Figure 1.*

*Figure 4. Chlorophyll content and fresh weight of the soybean first-leaf pair. Legend as in Figure 1. Plants treated for 4 hrs at 50 ± 5 pphm were assayed 24 and 48 hrs later.*

the first leaf pair (Fig. 4), fresh weight was not reduced by
ozonation at 6 or 18 days from sowing; but fresh weight specific
chlorophyll content was reduced by the treatment until the leaf
approached maturity as judged by constant fresh weight (about 16
days). Following ozonation at each of the four points on the
growth curve, purple specks which varied in density with leaf
age appeared uniformly over the leaf. These visible specks represented anthocyanin (optical density maximum at 527 nm in
acidic-methanol extracts), but not its overproduction since they
only became visible when the masking effect of chlorophyll was
removed. In addition to these visible injury sumptoms, larger
areas of bleached tissue appeared on the leaf pair following all
treatments except that given at 18 days. Active growth was required for visible expression of bleaching. No generalized
necrosis was observed, but neither bleaching nor anthocyanin-
speckling were repaired during incubation for a further 5 days.

As with the leaf pair, the fresh weight of the soybean
trifoliate leaf (Fig. 5) was not reduced in either very young
(10 days) or full size leaves (18 days); but it was reduced when
14-day seedlings were treated. The fresh weight specific
chlorophyll content was reduced following each treatment. This
damage was not overtly visible following ozonation at 10 days,
when the young trifoliate leaf is normally pale green, but it
was seen 24 hr after the 14-day treatment as speckle and then
became generalized bleaching with an additional 24 hr of incubation. The reduction in chlorophyll content after treatment of
18-day seedlings was never visibly manifest as bleaching except
for the presence of anthocyanin speckle. The second trifoliate
leaf, however, developed damage symptoms identical to those seen
in the treated 14-day first trifoliate leaf, including anthocyanin speckle. Each new trifoliate leaf would be expected to be
similarly responsive.

Radiolabelled Amino Acid Utilization Immediately Following
Ozonation. In order to distinguish early or primary events of
ozone damage from any later display we turned to radiolabelling
techniques, measuring the uptake of $^{14}$C-algal protein hydrolysate
into the soluble pool and its incorporation into protein in
tissue discs. Discs were taken from plants immediately following
in vivo ozonation, and were merely suspended in the labelling
solution to permit as normal as possible rates of amino acid uptake. All experiments were performed under conditions which
permitted linear uptake of radiolabel, and precautions were taken
to eliminate apparent free space label from consideration (see
Materials and Methods).

Damage Induced by Short-Term Ozone Exposure. Cucumber and
soybean seedlings of ages determined earlier to be overtly
sensitive to fixed dose and dose-rates of ozone (Table I) were
exposed to 45 ± 5 pphm ozone for periods up to 90 min. Some

treated plants were held for 24 hr after ozonation, and the usual
fresh weight and chlorophyll assays made. These data have been
given in Table I to show that damage symptoms depended in a non-
linear way upon length of time of exposure to ozone at a fixed
dose-rate, and were distinguishable after 15 min exposure.

Temperature-Dependence of Amino Acid Uptake in Soybean. The
uptake of amino acids by plants into the soluble pool can be an
energy-requiring process (52-54), as is the support of protein
synthesis. We wished to show the temperature-dependence of
influx of amino acids into soybean trifoliate leaf discs taken
from control and ozonated seedlings in order to confirm this for
the present system. Table II shows that amino acid influx during
30 min was reduced by the lowered temperature by about 70%, and
protein synthesis by about 85%, when labelling of control discs
proceeded at ice bath temperatures. Qualitatively similar reductions occurred in discs taken from ozonated (50 ± 5 pphm,
90 min) leaves and then labelled for 30 min. The ozone treatment
itself reduced the soluble pool size measured at 26°C by approximately 30%, and reduced incorporation by 40% in discs from the
soybean trifoliate leaf. When measured at ice bath temperatures,
however, label uptake was not reduced by ozonation, as would
be expected if the low temperature observations depended upon
the anatomical temperature-sensitivity of the soybean leaf.
Thus, we assumed that the system of amino acid uptake into the
soybean pools with which we were working was predominantly
energy-dependent.

Inhibition of Uptake into the Soluble Pool after Short-Term
Ozone Exposure; Non-Linearity of the Response. Figure 6 represents the uptake of $^{14}$C-labelled amino acids into the soluble
pool (see Materials and Methods) during 30 min incubation time
in discs taken immediately after the given length of ozone exposure, using seedlings from the same cultures as those represented in Table I. Thus, for the 15 min data in Figure 6, a
total of 55 min elapsed between the onset of ozonation and homogenization of the water-washed discs. Figure 7 shows that 15 min
of ozonation at 45 ± 5 pphm perturbed the amino acid incorpo-
rating system, but that no further reduction in rate was detected
as the length of ozone exposure was increased to 90 min. The
rate of uptake of label into the soluble pool (Fig. 6), however,
was affected in a non-linear, dose-dependent way for each organ
tested. Only the cucumber cotyledon showed increased permeability
following ozonation, which correlates in this tissue with the
visible anatomical changes that occurred during ozonation.

Immediate versus 24 hr Effects upon Amino Acid Utilization
Following Short-Term Ozone Exposure. To work with appreciable
differences between control and ozonated discs, the respective
seedlings were ozonated for 90 min at 45 ± 5 pphm ozone, and

Table II.  Temperature-sensitivity of amino acid utilization in soybean

| Treatment | Utilization of $^{14}$C-protein hydrolysate (cpm)* | | |
| | Temperature of labelling | | 1-2°C temp. as percent of 26°C temp. |
| | 26°C | 1-2°C | |
|---|---|---|---|
| Control    soluble pool   | 41,011 ± 1,597 | 13,215 ± 1,276 | 32 |
| Control    incorporated   | 17,744 ±   636 |  2,504 ±   185 | 14 |
| Ozonated   soluble pool   | 29,437 ± 1,104 | 14,640 ±   185 | 50 |
| Ozonated   incorporated   | 10,472 ±   464 |  4,030 ±   813 | 38 |

* 14-day-old soybean plants treated at 50 ± 5 pphm for 90 min, and discs taken immediately. The discs were exposed to 1.6 μc/ml $^{14}$C-labelled algal protein hydrolysate at either 26 or 1-2° as given in Materials and Methods.

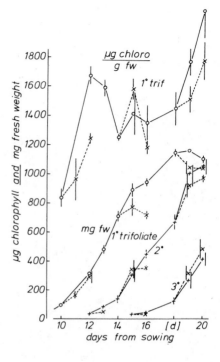

Figure 5.   Chlorophyll content and fresh weight of the soybean trifoliate leaf. Legend as in Figure 4.

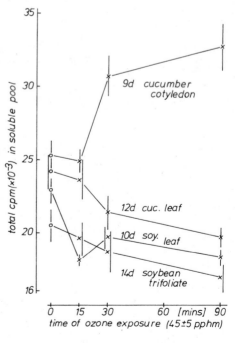

Figure 6.   Use of $^{14}$C-protein hydrolysate immediately following ozonation: soluble pools.

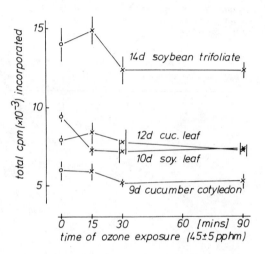

Figure 7. Use of ¹⁴C-protein hydrolysate immediately following ozonation: incorporation.

amino acid label utilization was measured in the excised discs
taken both immediately after treatment and 24 hr later
(Table III). The immediate increase in uptake into the soluble
pool of the cucumber cotyledon was not obvious, and there was a
greater immediate relative reduction in amino acid incorporation
due to the ozone treatment. After 24 hr incubation of treated
plants -- and when injury symptoms were visible -- uptake and
incorporation of labelled amino acids were greatly increased
relative to both the untreated controls and the time-zero treated
discs. Thus, an initial depression in both uptake (1 to 24%)
and incorporation (23 to 48%) was observed prior to the appearance of visible injury symptoms, followed by an absolute and
relative-to-controls increase in both uptake (7 to 32%) and
incorporation (8 to 78%) when damage symptoms were visible 24 hr
later. The treated tissue was, therefore, viable with respect
to $^{14}$C-amino acid utilization during the experiment, and the
observed increases in label uptake are taken to represent repair
processes of damaged tissue.

Charging of Amino Acyl-tRNA's Did not Influence the Ozone-
Induced Reduction in Soluble Pool Sizes. The possibility that
the observed reduction in amino acid pool sizes might have been
due to a reduction in charging of the amino acids to their
respective tRNA's was excluded in the following way using the
12-day cucumber leaf. Discs from plants treated with 47 ± 2 pphm
ozone for 30 min were labelled with $^{14}$C-protein hydrolysate for
30 min, and homogenized after cold tap water washing for 10 min
intervals up to 60 min. The pellet was washed with TCA/acetone
and digested with 0.1 M TRIS-HCl at pH 9 for 10 min at room
temperature, chilled, made 10% (w/v) TCA and centrifuged. Counts
in the supernatant were taken as the amount of label present in
charges tRNA. Ozone-treated plants did not differ from controls
in the incorporation of label into charged tRNA even while uptake into the soluble pool was lowered by about 40%. Total counts
present in the amino acyl-tRNA fraction were on the order of 1%
of counts in the soluble pool, and 4% of the counts incorporated
into protein; this was not changed in vivo during water-washing
of the labelled discs for up to 60 min before homogenization.

Pool-Loading Experiments and Possibility that Reduction in
Incorporation Reduces Influx into Soluble Pool. In order to
empirically eliminate the possibility that the observed reduction
in uptake of labelled amino acids into the soluble pool was an
indirect result of an inhibition of incorporation, it was
necessary to experimentally separate the influx and incorporation
systems in vivo. The operational compromise required to
accomplish this has been given by Sacher (55), and involves in
the present case measuring the rates of label incorporation in
control and treated tissues which have had their soluble pools
fully loaded with unlabelled amino acid. Thus, the label-specific

Table III. Immediate and delayed effects of ozonation upon amino acid utilization

| Tissue | Treatment | Total cpm 14C-amino acids/30 min | | | |
|---|---|---|---|---|---|
| | | Immediate | | 24 hr after treatment | |
| | | Soluble pool | Incorporated | Soluble pool | Incorporated |
| 9d Cucumber Cotyledon | Control | 31,066 ± 783 | 9,300 ± 546 | 27,671 ± 1,948 | 7,575 ± 634 |
| | Ozone | 30,806 ± 2,400 | 7,175 ± 839 | 36,537 ± 1,113 | 13,467 ± 536 |
| 12d Cucumber Leaf | Control | 24,582 ± 1,706 | 14,275 ± 952 | 27,786 ± 1,125 | 11,250 ± 966 |
| | Ozone | 18,586 ± 1,045 | 7,425 ± 1,131 | 35,347 ± 2,082 | 17,775 ± 1,040 |
| 10d Soybean Leaf | Control | 19,406 ± 1,121 | 10,800 ± 858 | 22,292 ± 574 | 7,500 ± 108 |
| | Ozone | 16,374 ± 737 | 7,724 ± 489 | 24,771 ± 693 | 9,200 ± 589 |
| 14d Soybean Trifoliate | Control | 19,513 ± 735 | 10,350 ± 813 | 21,514 ± 771 | 9,367 ± 982 |
| | Ozone | 15,982 ± 807 | 7,300 ± 803 | 23,033 ± 694 | 10,100 ± 567 |

activity is made low in each soluble pool, and the rates of label
incorporation will then minimally depend upon the label-specific
activity of the pools of each.  This is because it is expected
that the pool specific activities would be approximately equal,
and incorporation rates would maximally depend upon the incorporation process itself.  If the incorporation rates do not differ
under these conditions, then the original observation that the
ozone treatment has altered soluble pool sized may be argued
exclusively.

    This kind of experiment was performed in an attempt to show
that reduction in labelled amino acid uptake into the soluble
pool, observed after as little as 15 min of ozonation (Fig. 6),
could be an early or primary event in ozone damage to the soybean
trifoliate leaf.  It was necessary in this case for the pools to
be loaded with unlabelled casein hydrolysate after treatment of
the plant, in order to maintain in vivo ozone treatment and because it has been reported that high endogenous amino acid levels
can confer a degree of resistance to ozone damage (14, 15).
Plants were ozonated at 50 ± 5 pphm for 90 min; freshly cut discs
were shaken for 60 min in 50 mg/100 ml casein hydrolysate or in
distilled water; $^{14}C$-protein hydrolysate was added and the discs
were incubated for 60 min in the light; discs were then washed in
running cold tap water for 10 min and homogenized in TCA/acetone.
Thus, a total of 3 3/4 hr elapsed after the onset of ozonation
and before homogenization, as opposed to 55 min in the case of
15 min ozonation followed by 30 min labelling time plus 10 min
washing time (Fig. 6).

    Table IV shows typical results of the pool overloading
experiments in the soybean trifoliate leaf.  Visible damage to
ozonated plants after 24 hr incubation was, as usual, taken as
necessary for an experiment to be valid.  Clearly, incorporation
of label into protein was reduced where casein hydrolysate had
been used in control and treated discs to overload the soluble
pools.  Just as clearly, total counts present in the soluble
pool of treated tissue were reduced by ozonation whatever the
after-treatment.  The reduction of label in the soluble pool in
control and ozonated discs due to casein treatment was approximately 30% in each case; the reduction of label incorporation
due to ozone was about 36% in the case of water treatment and
about 14% in the case of casein treatment.

    We conclude that by 3 3/4 hr after the onset of ozonation
the inhibition of protein synthesis has become a factor in the
overall physiological disturbance in the soybean trifoliate leaf.
However, this was not necessarily the case for the earlier
(55-100 min) events described in Figure 6.  That data showed that
protein synthesis was depressed by as little as 15 min of ozone
exposure, but not depressed further as ozone exposure was
lengthened to 90 min.  What we have been unable to show by the
overloading experiment is the exclusivity, immediately following
ozonation for brief periods, of the inhibition of amino acid

influx. This is true despite the fact that 3 3/4 hr after the
onset of ozonation it was still possible to observe the reduced
influx of label into the soluble pools when discs were pre-loaded
with casein hydrolysate (about 30% reduction, Table IV). We have
been unable thus far to design the overloading experiment in a
way which would greatly shorten the elapsed time between ozonation and assay.

Table IV.  Distinction between amino acid influx and
incorporation

| | Utilization of $^{14}C$-protein hydrolysate (cpm)* Total cpm, $^{14}C$-amino acids | |
|---|---|---|
| Treatment | Soluble pool | Incorporated |
| Control + water | 342,441 ± 42,321 | 92,830 ± 4,950 |
| Ozonated + water | 256,391 ± 3,940 | 59,320 ± 2,378 |
| Control + casein | 241,164 ± 13,040 | 41,290 ± 889 |
| Ozonated + casein | 182,468 ± 11,076 | 35,455 ± 1,562 |

*Experimental details given in text and Materials and Methods.

## Conclusions

a. The present work confirms that active growth of the
organs of cucumber and soybean seedlings is required to detect
ozone-induced inhibition of chlorophyll accumulation. During
active growth of the respective organs there were no periods of
differential ozone-sensitivity (Fig. 1,2,4,5). Short exposures
to ozone (15 min) were followed after 24 hr by an increase in
fresh weight per organ in all cases but the 10-day soybean first
leaf (Table I). The response to ozone was always non-linear by
whatever damage parameter used. Organs which repeat during
development (cucumber leaf and soybean trifoliate leaf) repeat
the pattern of their original sensitivity to moderate levels
of ozone exposure.

b. The temperature-dependent (Table II) uptake of amino
acids into the soluble pools (Fig. 6) of organ discs is sensitive
to short exposures to ozone, as is the incorporation of amino
acids into protein (Fig. 7). Incorporation appears to be reduced
by a fixed increment during the first 30 min of ozone treatment,
and then no further (Fig. 7), while uptake into the soluble pool
is reduced (except in cucumber cotyledons) in a non-linear way
throughout 90 min exposure periods (Fig. 6).

The viability of ozonated seedlings was attested to by the
fact that after visible injury symptoms had developed in treated

plants, the uptake and incorporation of amino acids into discs were greatly increased in relation to both untreated controls and the time-zero treatment discs (Table III).

c. The observed rapid reduction in amino acid pool sizes in all but the cucumber cotyledon was not due to a reduction in the tissue's capacity to charge the respective tRNA's as measured in the 12-day cucumber leaf disc. An alternative possibility, that the reduction in amino acid incorporation into protein was indirectly responsible for rapidly lowered pool sizes, could not be expressly excluded when 3 3/4 hr had elapsed between the onset of ozonation and assay. The exclusivity of ozone-inhibition of amino acid uptake into soluble pools cannot, therefore, be argued 3 3/4 hr after the onset of ozonation. As Figure 7 shows, however, there does not appear to be any further reduction after 30 min of ozone treatment in protein synthesis (70 min elapsed time) and we conclude that it is this time differential (70 vs. 220 min elapsed time) which has permitted protein synthesis to become measurably inhibited.

d. The data presented here provide direct evidence for the ozone-induced rapid inhibition of the potential for amino acid uptake into the soluble pool of cucumber and soybean seedlings. This evidence implies that an early or primary effect of ozone damage to these seedlings is the interruption of transport processes at the level of the plasma membrane, and supports the interpretations of Evans and Ting (45) with Phaseolus, and Nobel and Wang (46) with Pisum to this effect using quite different assay techniques. The interesting question, of course, deals with the way(s) in which ozone might react with the plasma membrane rapidly (Fig. 6), cooperatively (Fig. 3 and 6, Table I) and selectively (Fig. 6, cucumber cotyledon vs. cucumber first leaf) to affect it functionally (cf. 6,22,25).

Acknowledgements: This study was supported in part by the U. S. Environmental Protection Agency, grant No. 1-ROL-APO-1223. This report is journal paper No. 5593 of the Purdue Agriculture Experiment Station. The secretarial assistance of Ms. Sian Frick is gratefully acknowledged.

## Literature Cited

1.  Daines, R. H.  In "Current Topics in Plant Science", ed.
    J. E. Gunckel.  p. 436.  Academic Press, New York. (1969).
2.  Hill, A. C., M. R. Pack, M. Treshow, R. J. Downs and L. G.
    Transtrum.  Phytopathology (1961) 51:356.
3.  Howell, R. K. and D. F. Kremer.  J. Environ. Qual. (1972)
    1:94.
4.  Reinert, R. A., D. T. Tingey and H. B. Carter.  J. Amer.
    Soc. Hort. Sci.  (1972) 97:711.
5.  Treshow, M.  Phytopathology (1968) 58:1108.
6.  Treshow, M.  Environmental Pollution (1970) 1:155.
7.  Anon.  "Losses in Agriculture" Agr. Res. Service, U.S.D.A.
    Agri. Handbook 291, 1965.
8.  Anon.  "Air Quality Criteria for Photochemical Oxidants"
    U. S. Dept. Health, Education and Welfare, Nat'l. Air Pollu-
    tion Control Association Publication  Ap-63.  (1970).
9.  Stern, A. C.  "Air Pollution"  Academic Press, New York.
    (1968).
10. Dugger, W. M. and I. P. Ting.  Ann. Rev. Plant Physiol.
    (1970) 21:215.
11. Heck, W. W.  Ann. Rev. Phytopathology (1968) 6:165.
12. Menser, H. A., H. E. Heggestad and O. E. Street.  Phyto-
    pathology (1963) 53:1304.
13. Mudd, J. B., R. Leavitt, A. Ongun and T. T. McManus.  Atmos.
    Environ. (1969) 3:669.
14. Ting, I. P. and S. K. Mukerji, Amer. J. Bot. (1971) 58:497.
15. Tomlinson, H. and S. Rich.  Phytopathology (1967) 57:972.
16. Adedipe, N. O., G. Hofstra and D. P. Ormrod.  Can J. Bot.
    (1972) 50:1789.
17. Craker, L. E.  Can. J. Bot. (1971) 49:1411.
18. Lee, T. T.  Can. J. Bot. (1966) 44:487.
19. Menser, H. A. and G. N. Hodges.  Tobacco Science (1967)
    6:151.
20. Dugger, W. M., O. C. Taylor, E. Cardiff and C. R. Thompson.
    Proc. Amer. Soc. Hort. Sci.  (1962) 81:304.
21. Lee, T. T.  Can. J. Bot. (1965) 43:677.
22. Mudd, J. B., T. T. McManus, A. Ongun and T. E. McCullogh.
    Plant Physiol. (1971) 48:335.
23. Goldstein, B. D. and O. T. Balchum.  Proc. Soc. Exp. Biol.
    Med.  (1967) 126:356.
24. Tomlinson, H. and S. Rich.  Phytopathology (1969) 59:1284.
25. Swanson, E. S., W. W. Thomson and J. B. Mudd.  Can. J. Bot.
    (1973) 51:1213.
26. Scott, D. B. M. and E. C. Lesher.  J. Bacteriol. (1963)
    85:567.
27. Siegel, S. M.  Plant Physiol. (1962) 37:261.
28. Tomlinson, H. and S. Rich.  Phytopathology (1968) 58:808.
29. Freebairn, H. T.  J. Appl. Nutrition (1959) 12:1.
30. Barnes, R. L.  Environmental Pollution (1972) 3:133.

31. Erickson, L. C. and R. T. Wedding. Amer. J. Bot. (1956) 43:32.
32. Lee. T. T. Plant Physiol. (1967) 42:691.
33. Pell, E. J. and E. Brennan. Plant Physiol. (1973) 51:378.
34. Wilkinson, T. G. and R. L. Barnes. Can. J. Bot. (1973) 51:1573.
35. Leffler, H. R., L. B. Baggett and J. H. Cherry. Plant Physiol. (1972) 49 Suppl.:16.
36. Tingey, D. T., R. C. Fites and C. Wickliff. Physiol. Plant. (1973) 29:33.
37. Ordin, L., M. A. Hall and J. I. Kindinger. Arch. Environ. Health 18:623.
38. Chang, C. W. Phytochemistry (1972) 10:2863.
39. Ledbetter, M. C., P. O. Zimmerman and A. E. Hitchcock. Contrib. Boyce Thompson Inst. (1959) 20:275.
40. Evans, L. S. and P. R. Miller. Amer. J. Bot. (1972) 59:297.
41. Thomson, W. W., W. M. Dugger and R. L. Palmer. Can. J. Bot. (1966) 44:1677.
42. Adedipe, N. O., H. Khatamian and D. P. Ormrod. Z. Pflanzenphysiol. (1973) 68:323.
43. Fletcher, R. A., N. O. Adedipe and D. P. Ormrod. Can. J. Bot. (1972) 50:2389.
44. Ting, I. P. and W. M. Dugger. J. Air Pollution Control Assoc. (1968) 18:810.
45. Evans, L. S. and I. P. Ting. Amer. J. Bot. (1973) 60:155.
46. Nobel, P. S. and C. T. Wang. Arch. Biochem. Biophys. (1973) 157:388.
47. Giese, A. C. and E. Christensen. Physiol. Zool. (1954) 27:101.
48. Rich, S. Ann. Rev. Phytopathol. (1964) 2:253.
49. Frick, H. and H. Mohr. Planta (1973) 109:281.
50. Arnon, D. I. Plant Physiol. (1949) 24:1.
51. MacRobbie, E. A. C. Ann. Rev. Plant Physiol. (1971) 22:75.
52. Birt, L. M. and F. J. R. Hird. Biochem. J. (1958) 70:286.
53. Francki, R. I. B., M. Zaitlin and R. Jensen. Plant Physiol. (1971) 48:14.
54. Cheung, Y. N. S. and P. S. Nobel. Plant Physiol. (1973) 52:633.
55. Sacher, J. A. Symp. Soc. Exp. Biol. (1967) 21:269.

# INDEX

149